Inhuman Thoughts

Inhuman Thoughts

Philosophical Explorations of Posthumanity

Asher Seidel

LEXINGTON BOOKS

A division of

ROWMAN & LITTLEFIELD PUBLISHERS, INC.

Lanham • Boulder • New York • Toronto • Plymouth, UK

LEXINGTON BOOKS

A division of Rowman & Littlefield Publishers, Inc.
A wholly owned subsidiary of The Rowman & Littlefield Publishing Group, Inc.
4501 Forbes Boulevard, Suite 200
Lanham, MD 20706

Estover Road
Plymouth PL6 7PY
United Kingdom

British Library Cataloguing in Publication Information Available

Library of Congress Cataloging-in-Publication Data

Seidel, Asher, 1943–
Inhuman thoughts : philosophical explorations of posthumanity / Asher Seidel.
p. cm.
Includes bibliographical references and index.
ISBN-13: 978-0-7391-2328-7 (cloth : alk. paper)
ISBN-10: 0-7391-2328-9 (cloth : alk. paper)
1. Human beings. 2. Human evolution. 3. Humanity. I. Title.
BD450.S3854 2008
128—dc22

2007052128

Printed in the United States of America

∞™ The paper used in this publication meets the minimum requirements of American National Standard for Information Sciences—Permanence of Paper for Printed Library Materials, ANSI/NISO Z39.48-1992.

Contents

Preface

My overall thesis is that human beings, as presently constituted, will likely alter their nature profoundly in the sometime future. By "their nature" I intend their physiological, psychological, and sociological circumstances. By "the sometime future" I mean something further than the foreseeable future. How much further I cannot say; however, it might aid intuition to choose reference points such as one thousand years, twenty thousand years, one hundred thousand years (it should be noted that, in our present biological form, we have existed as a species for approximately one hundred thousand years). It is part of my overall thesis that there are various changes for the better that are worthy of our consideration.

I do not spend considerable time defending this thesis. Such defense as I present it is distributed throughout the work, with some focus in the concluding chapter. Should the reader desire reason for entertaining my thesis, the reader is urged to peruse initially the final chapter.

In another manner, the thesis is supported throughout the work by the elaboration of my vision of the movement from the improved-but-still-human to the post-human. The logic of this support is direct: if something is better, there is prima facie reason for preferring it. Much of what I present is in service of showing that the described artificial-evolutionary path suits humanity better than remaining as the species we have been.

There is scant philosophical literature on the topics I investigate. I know of only one philosophical book in the area, and that is an introductory text.[1] There are other works addressed to the general topic, but they are not by philosophers. Typically, the authors are journalists,[2] or specialists in one or more of the areas covered by the journalists.[3] There is additionally a technical literature on the biology of extended lifespan.[4] Hence, I cannot offer much comment on the philosophical literature which directly addresses the topic. Anyone philosophically inclined to explore this area is on his or her own.

Many historical philosophers spoke to the topic of human betterment. To name a few: Plato, Spinoza, Marx, Nietzsche. Any philosopher offering a normative theory is implicitly, if not explicitly, on the topic. Nevertheless, thoughts on the betterment of humans as humans is one thing; thoughts on the betterment of humans through transcendence of the biological human is another, the latter typically reserved in the tradition for either theology or science fiction. Only recently have other voices, such as those referenced in the preceding paragraph, emerged.

I do not proceed to a confident conclusion (and various sub-conclusions are likewise presented as merely probable or possible) because the matters discussed herein are not the sort about which one ought to be reasonably confident. This is primarily a speculative work, with elaboration of such speculative-theoretical notions as scientific ethics, immortality, parallel consciousness, extended visual perception, and nonsocial post-human life. Such topics are in large part beyond the pale of more standard philosophical disputations. Caution is proper.

In the course of developing these thoughts, I have been aided by discussions with various of my colleagues in the Philosophy Department at Miami University. I have also been helped by such feedback as I received in the presentation of aspects of this work at several gatherings of the Society for the Contemporary Assessment of Platonism, which sessions occurred at conventions of the Pacific Division of the American Philosophical Association. An anonymous reviewer for Lexington Books provided helpful suggestions. My editor, Patrick Dillon, gave me welcome support.

Two of the chapters, "On Human Improvement" and "Facing Immortality," have been published, with slight differences in the *International Journal of Applied Philosophy* (1999 and 2005), and they appear here with the kind permission of the editor of that journal.

NOTES

1. Brian Cooney, *Posthumanity: Thinking Philosophically about the Future* (Lanham, Md.: Rowman & Littlefield, 2004).

2. For example, Joel Garreau, *Radical Evolution: The Promise and Peril of Enhancing Our Minds, Our Bodies—and What It Means to Be Human* (New York: Doubleday, 2005); Ramez Naam, *More Than Human: Embracing the Promise of Biological Enhancement* (New York: Broadway Books, 2005); and Simon Young, *Designer Evolution: A Transhumanist Manifesto* (Amherst, N.Y.: Prometheus, 2006). I believe

the strongest counter to my overall thesis is presented by Francis Fukuyama, *Our Posthuman Future: Consequences of the Biotechnology Revolution* (New York: Farrar, Straus and Giroux, 2002).

3. See, for example, Hans Moravec, *Robot: Mere Machine to Transcendent Mind* (New York: Oxford University Press, 2000) and Ray Kurzweil, *The Age of Spiritual Machines: When Computers Exceed Human Intelligence* (New York: Viking Penguin, 1999).

4. Stanley Shostak, *Becoming Immortal: Combining Cloning and Stem-Cell Therapy* (Albany: State University of New York Press, 2002).

Chapter One

Introduction

The title is in a manner misleading, in a manner accurate. The essays contained herein are not expositions or explorations of inhumanity, monstrous behavior such as extreme cruelty, or unfeelingness in the face of the suffering of others. They are rather, to varying extents, in various ways, explorations of either enhanced humanity or non-humanity, the latter seen as an evolvement from humanity. Titling this work "Non-human Thoughts" would be in this regard more accurate, but alas less eye-catching.

Descriptive of the author of these pieces, however, the modifier "inhuman" carries the disapprobation reserved for one whose thoughts lead where they ought not, much as Stevenson's Dr. Jekyll traveled forbidden terrain. Nor should it be allowed that no moral boundaries exist for thought. There are questions one should not ask. If one allows one's imagination a momentarily ghastly turn, one will have little difficulty framing such questions. It might be thought that these studies run wide of the acceptable, inasmuch as they propose a movement from the human to the non-human. All the more objectionable that this is held a positive movement.

Yet there have been those philosophers who have written about transcending the human, or at least the all-too-human. The latter sort of "overcoming," often associated with Nietzsche's Zarathustra, is not the object of examination here. But Nietzsche's call for a profound transformation in human thought and behavior is not the sole philosophical plea for something other than ordinary human nature. Religious philosophers often speak of otherworldly transcendence of the human. Plato, in perhaps his most extreme statement in this regard, argues in the *Phaedo* that embodied human life on earth is incompatible with full realization of knowledge, capable of a soul unencumbered with bodily perceptions and desires.

Such searching for positive value human-to-non-human transformation, whether by philosophers, theologians, or science-fiction writers, is typically labeled "bourgeois," or "escapist," or "utopian" by those opposed to the idea of such search.[1] To some extent, these charges are first addressed in the second essay in this collection: "On Human Improvement." But in that essay the assumption is that humans are improved, and not transformed into other-than-human entities. This last rejoinder raises the question of the nature and significance of the human-non-human boundary. Given sufficient latitude based on the supposed ambiguity of such a boundary, whatever countervailing force against charges of "escapism," etc., is manifested by the remarks in "On Human Improvement" arguably carries over to protect the more extended departures from ordinary human nature.

This speculated carryover has intuitive, if vague, limitations. As with many cases of vague boundary, there are positions sufficiently removed from borderline areas to disallow the appeal to vagueness. Some of the essays in this collection will occupy such positions, in virtue of which they are open to the charges given above, should the only defense against such charges be that they are addressed in the essay "On Human Improvement." And the boldness of the assumption of non-humanity in these essays opens them to another charge: that they are fantastic in the sense that, whatever the far future holds, current attempts at detailing such a future are bound to look ridiculous in hindsight. Recall the visions of cinema serials in the 1930s as to the rocketship future of humankind for a ludicrous vision of the shape of things to come. Within the space of fifty years or less, most science-fiction cinema and writing acquires a dated cast. Even visionaries as celebrated as Jules Verne are honored more for the generality of their occasionally successful predictions than for the accuracy of detail in them.

The foregoing criticisms constitute an a priori case against speculative extravagancies of the sort to follow. Much of this criticism must be allowed. It is doubtless more important that much philosophical labor be devoted to current and near-future problems of human affairs, than to speculating on the far future. And speculating on the far future in any but the most general manner is likely to be wide of the mark. Yet I urge that one reserve judgment until what has been done here is examined. I further submit that the occasional speculative flight made with the intention of improvement of the current human situation, even such improvement as cannot be realized in the present or near future, ought not be condemned outright as bourgeois, escapist, or utopian (although to some extent it is all these). Granting that such speculative flights ought not be universally prescribed for all moral philosophical writing, one might yet offer that some universalization ought take account of probable or expected distribution of effort. That is, the "what if everyone did that" criterion of rightness seems at times best applied as "what if everyone did as they are doing in the proportions they are doing it." Some people farm. What if everyone farmed? Universal farming, at the expense of all other forms of productive activity, is obviously self-negating in brief time. What if everyone engaged in full time artistic activity? And so on. Farmers plead for their activity on obvious grounds of social usefulness. Artists plead for their activity on somewhat less firm ground; futurist speculators, even less. These types of activity are standardly supported, more or less, by proportions of social resources commensurate with their social usefulness.

Is there any social usefulness to any form of deep-future speculation? In response, one might appropriate whatever social usefulness attends aesthetic ventures and declare deep-future speculation an aesthetic venture. This move might cover most science-fiction endeavors, but it is unlikely to satisfy those doubting the social usefulness of philosophical speculation regarding the deep future. One might have similar doubts about the social usefulness of much of the traditional and current metaphysics and epistemology. What social purpose is served by debates regarding competing accounts of mundane material-object identity over time, or the internal or external nature of justification of epistemic claims? And while the social-epistemological movement is on prima facie better grounds here, the promise of socially significant results from investigations of the socially situated character of epistemic claims arguably looks better at a distance than a closer view of these often trend-influenced discussions affords.

As with a significant amount of philosophy, philosophical speculation regarding the far future is guilty of the Marxist charge of bourgeois thought. Unlike metaphysics and epistemology, it runs a good chance of being guilty of the charge of being utopian, at least on the assumption that much of the sort of speculation to be manifested here is unlikely to be realized in the foreseeable future. The charge of being escapist is perhaps avoidable, especially if this charge is intended in the sense that the author is in some psychological manner seeking to avoid genuine concerns. Anyone so charged has all the rationalizing defenses open to reply that much of their time is consumed in socially relevant work, that they cannot be expected to be constantly vectored to social concerns (they must sleep, for example), and so forth. To the extent that these responses appear mere rationalizations, however, the escapist charge lends support to the charge of bourgeois thinking.

Held against these charges, why ought not philosophical speculation regarding the deep future be dismissed outright on grounds of uselessness? The answer is as obvious as it is age-old. People speculate, and at times enjoy sharing speculation. Even as art often claims an end-in-itself relief from demands of social utility, so speculative thought claims such relief. Those societies seeking to maximize socially valuable productive activity may, with good reason, de-emphasize the aesthetic life, and likewise the philosophically contemplative life. Those societies in which problems of production and distribution seem more due to political circumstances than due to labor shortage, however, may justifiably consider various forms of contemplation allowable ends-in-themselves. What remains at issue includes the extent to which any society is so situated with regard to resources, the obligations to societies not so situated placed on societies that are so situated, and the possibly socially corrosive effects of countenancing nonproductive activity even in a society not requiring the full time productive activity of all its members.

We must acknowledge the challenge that far future speculation is likely to be ridiculously inaccurate to what eventually comes to pass, such inaccuracy being proportionate to the extent that the speculation descends to specifics. Since any lengthy exploration of future possibilities will exhibit some degree of detail, the detail should be approached with the attitude of good humor. Of course, specificity and generality are, to a significant extent, relative notions, and admit of varying application. It would doubtless be an error of specificity to detail the appliances in the kitchen of 10,000 years hence, or even to presume something such as a kitchen at that time. If biological humans are the subjects of discussion, however, it may not be amiss to speak generally of nutrition, nor to speculate regarding general alternatives such as synthesized foodstuffs in lieu of agriculturally derived sustenance.

With one or two exceptions, the essays in this collection are well into the "absurdity zone." Speculation will concern humans having willfully evolved into non-humans. Such evolution will be considered in three manners: biologically, psychologically, and sociologically. Speculation on these changes will, at times, involve a degree of detail sufficient to raise skepticism as to the basis for suggesting the possibility. In most, if not all, these relatively detailed speculations, detailed suggestions will be just that—suggestions. If it is imagined that people have evolved into entities constructed of silicon and titanium, for example, less emphasis will be placed on the particularities of silicon and titanium—which are given simply to supply more specific values for the blank spaces created by the general suggestion of evolution from the carbon-based to the non-carbon-based, than on the general suggestion itself.

The reader will hopefully be startled by the suggestion that humans evolve into entities primarily composed of silicon and titanium. The reader will perhaps be temporarily relieved to hear that this suggestion will not be presented in the forthcoming text. I have no substantial basis for considering titanium and silicon to be candidate materials for human-to-non-human evolution. I have scant knowledge of the properties of these elements and of other such candidate materials. This lack of knowledge does not permit reasonable speculation on the chemical nature of what our deep-future selves might be. However, the reader is advised not to be too relieved. The second essay in this collection considers artificial changes to humans which, although not altering their intuitively human status, do involve such possibilities as brain implants. The later essays will suggest considerably more profound alterations.

There exist currently various prostheses such as artificial limbs, hearts, eyes (in the form of communicative sensors), and ears. Imagine a skin-enclosed human torso whose artificial extremities are such that they respond to central nervous system impulses much as do standard extremities. Imagine that major skeletal components have been replaced by artificial implants, and that bone marrow function (due to significant reduction of bone marrow) is performed in some artificial manner. Major organs such as heart, lungs, and liver are similarly artificial. Such an imagined human is but a step away from current biotechnological possibility. Blood still courses through the veins of this imagined person, the person's brain still functions much the same, the person ingests similar food, and has similar basic and higher "drives." If we speculate on further alterations, at what point does our imagined entity cease to be human?

I suspect that there will not be universal agreement on the answer to this question, and further, that many will find this question difficult, if not impossible, to answer. The question was posed lightly. The example will not be pressed here, but is presented as a forewarning of the plausibility of the speculations to follow—speculations which appear implausible when put in the "silicon and titanium" form as above, but may appear less so when given in another manner.

As a focusing aid, I propose the following: it is not the material composition of the entity that founds the human/non-human distinction, but rather various psychological and sociological departures from current and foreseeable (and [pre]historical) possibilities of human psychological and sociological experience. By "material composition" I mean not merely the result of a chemical assay, but the overall biological configuration in its highs and lows that distinguishes modern humans, homo sapiens sapiens, from other biological entities. One can imagine entities having the same external dimensions as modern humans, with the same basic needs and drives as modern humans (food, moderate climate, sex, childrearing, etc.), yet composed of different material than modern humans, and with different internal configurations. That is, one can imagine this if one imagines various changes in laws of nature so as to permit a radically different biology, or perhaps non-biology, to result in the various needs, drives, behavior, vulnerabilities (e.g., to pathogenic sicknesses) of humans. Rather, perhaps this can be imagined. Will such entities fear strokes, heart attacks, testicular or breast cancer? Obviously, the more detailed the imagination, the more inventiveness will be required to vivify the imagined entities. Alternatively, one can attempt the more limited task of imagining entities of radically different material composition than modern humans, duplicating humans in some area of activity heretofore restricted to biological humans. In this regard, various aspects of human cognitive behavior come readily to mind as candidates for duplication.

If the laws of nature are not ignored, or not selectively altered, then it seems unlikely that entities significantly similar to current humans (including the bodily concerns of current humans) will evince significantly altered material composition. From which it follows that entities of significantly altered material composition, in the sense of "material composition" given above, will have notable departures from current humans. While these equivalences do not necessitate that entities of significantly different psychological and social characteristics will depart significantly from the biological characteristics of current humans, such departure is suggested by the correlation of changes in composition with altered status vis-à-vis human/non-human.

Whatever the final judgment on the matter of correlation of biology with psychology and sociology, the following changes will be explored in the chapters succeeding the earlier chapters, "Revolutionary Ethics" and "On Human Improvement." First, in "Facing Immortality," I will consider the possibility and implications of a vastly extended lifespan including, as a final case, a lifespan with no limitations. Next, in "Parallel Consciousness" I will consider the possibility indicated by the title. In "Mindful Seeing," I suggest an overcoming of the barrier between concentrated thinking and simultaneous sense perception. In "Alone and Without Love" I contemplate a de-sexed society of entities evolved from humans, which is to say a society of once-humans (i.e., entities with a memory of having evolved, continuously or discontinuously, from humans) who no longer regard each other with sexual desire and who no longer have need of each other's company.

Such considerations as these seem best done, if at all, by science-fiction writers, futurists, future-oriented sociologists, and so forth. Until recently, natural scientists have been reluctant to speculate on such deep-future possibilities. One might question the role of philosophy in such speculation. To some extent, this question is confronted in the first two essays. While the speculations in "On Human Improvement" are confined to the foreseeable future, and for that reason have less of a science-fiction cast to them, they are nonetheless steps along the road to deep-future speculations. And the perennial philosophical questions of life's goals, and the manner in which to live one's life, in both prudential and moral senses, as well as related philosophical questions of sociopolitical organization, will be given fresh meaning when placed in these speculated contexts.

In some of these contexts, some of these questions apparently have no meaning at all. For instance, moral concerns seemingly do not arise in a context where once-moral agents no longer have need of each other's personal wealth, sex, or company. But such postulated entities might yet have the capacity, and the inclination (or tendency) to harm each other, so even given such extreme independence questions of morality may be applicable. And the prudential question remains as to the most self-beneficial manner in which such entities should exist.

To begin, let us consider the foreseeable future. Of course, the near future is not wholly foreseeable. There are catastrophic scenarios, both natural and human caused, that result in radical devolution, or even elimination, of humankind. But this book is written in an optimistic spirit. All the speculations concern possibilities thought to be positive. The supposition is that, given a state of affairs involving humans or once-humans, it is in some respects possible to posit an improvement of the human or once-human entities considered in that state of affairs. Among the questions raised by this supposition are (1) the nature of the judgment that one circumstance is an improvement on another; and (2) the relation of the judgment to its time of utterance (is what is deemed a future improvement when viewed in current circumstances necessarily an improvement when viewed in future circumstances?). These questions will be discussed at some length in the following chapters.

What is intended here by "the foreseeable future"? Simply put, it is those later moments of speculated time at which we have not ceased to be biologically, psychologically, or sociologically human. We can still recognize ourselves when positioned at these moments, although our capacities may be somewhat altered. This is admittedly vague, and somewhat misleading. Am I "recognizing" myself if I imagine myself on the surface of Pluto with no additional life-support system than the clothes I am now wearing? Am I recognizing myself if I imagine myself able to read a complex novel in two minutes, with a comprehension of what I have read equal to my current comprehension on completion of the novel?

At least some science fiction seems constrained by the demands of such self-recognition, in that in much of the genre human beings are placed in vastly altered technological circumstances with characteristics of the reader's friends and neighbors. There is much historical support for this constraint. We are not greatly different from Plato's friends and neighbors, despite our vastly altered technological circumstances. Hence, we are part of Plato's foreseeable future—given my use of the term. Plato's vision of human possibilities, as evinced in the *Republic*, was constrained by possibilities of production and general lack of scientific knowledge. Having expanded these possibilities, we are arguably capable of a psychological and sociological transformation that would not have been recognizable to Plato. Some visionaries of the nineteenth century and beyond have elaborated various possibilities of such transformations, but for the most part these transformations have not been realized.

I am not bounding speculation on future possibilities to foreseeable scientific and technological advances. In the concluding chapter I will suggest that the non-foreseeable future, as I intend the notion, may be closer than we typically think.

I want to say a few words regarding the motivation of this work and the influences occasionally noted therein. As I reflect on what I have thought and written, I am somewhat surprised at the manner in which this project took shape. What follows may be regarded as sadly mistaken, even pathological, yet it was arrived at honestly. I was not initially aware of the implications of my thoughts, and was at times startled as some of them became apparent to me. Rather than avoiding the more radical results (e.g., the implication of childlessness in "Facing Immortality"), I put them forth as they occurred to me. Tampering with human nature, even in thought experiment, is a fearsome business.

The book arose from reflection on my essay "On Human Improvement." That essay was conceived one day as I was teaching symbolic logic, observing my students. I had given them an in-class problem and watched them thinking and writing. The problem was more a matter of calculation than intuiting, say, the proof of a metatheorem. The calculation demanded chaining results of various routines with which the students were familiar. The routines themselves were thought intensive, rather than automatic in the way truth-tabulation is automatic, and the entire process took the average student a bit of time, the better student somewhat less time.

It struck me that the manner in which they struggled with the problem was typical of virtually any one of our species who would be confronted with this problem (the specifics of which I do not recall) at that point in their learning. Since there was an overall routine to be followed and a definite answer, I could imagine doing the problem quicker. Indeed, I could do the problem significantly faster than any of my students, due to my familiarity with the operations. In short, I had a sudden awareness of human limitations. There are obviously other ways of coming to this awareness. Had I thought about the time most humans require to solve a problem in long multiplication, compared to the time required by a handheld calculator (I am aware that at one level of description the handheld calculator uses a markedly different routine than the human calculator), I might have had a similar inspiration. Whatever might have been, once I had written "On Human Improvement" the seed of the remainder of the book was in mind.

As I thought about enhanced cognition, I recalled a paper I had written earlier on parallel conscious thinking. It seemed sufficiently related to the topic of human improvement to allow its being reworked and incorporated into the project. Considerations of enhanced cognitive capability led to considerations of extended lifetime, and by the time I was finished writing "Facing Immortality" I realized that I was near, if not over, the boundary separating human from non-human. Having gone that far, it was not difficult to push further, as is done in "Mindful Seeing" and "Alone and Without Love." The realization that I could not avoid ethical considerations led to the composition of "Revolutionary Ethics," less a discussion of traditional ethics than a hopeful call for something not yet accomplished.

To the extent that this work proceeds from recognizable influence, Plato is the most apparent such influence, whatever degree my thoughts depart from his. There are other influences, less by their overall thought than various specific thoughts they have articulated. Hume is cited on a few occasions in this regard. Some of Marx's thoughts are in the background, although he would likely disown any association with this work. Some speculations of the contemporary philosopher Paul Churchland initiated thoughts. With regard to Churchland I want to share my delight in the following. In "On Human Improvement" I reference Churchland's speculation regarding "tricking" neurons into active attachment with silicon devices. I have since been informed that such is no longer speculation.[2] It has been accomplished.

NOTES

1. There are other authors, some with either proper scientific credentials or credible acquaintance with relevant literature, whose speculations in this area are not of the science-fiction variety. See Joel Garreau, *Radical Evolution: The Promise and Peril of Enhancing Our Minds, Our Bodies – and What It Means to Be Human* (New York: Doubleday, 2005); Ramez Naam, *More Than Human: Embracing the Promise of Biological Enhancement* (New York: Broadway Books, 2005); Keith Stanovich, *The Robot's Rebellion* (Chicago: University of Chicago Press, 2004); and SimonYoung, *Designer Evolution: A Transhumanist Manifesto* (Amherst, N.Y.: Prometheus, 2006).

2. See the work done by Peter Fromherz and associates at the Max Planck Institute for biochemistry: Peter Fromherz, and A. Offenhäusser, T. Vetter, J. Weis, "A Neuron-Silicon Junction: A Retzius Cell of the Leech on An Insulated-Gate Field-Effect Transistor," *Science* 252, (1991): 1290–93.

Chapter Two

Revolutionary Ethics

What is the point of ethics? To ask this question is to invite the response that the philosophical discipline of ethics has various goals. At times ethics is said to be descriptive, at other times prescriptive. Yet from among these goals there emerges the fundamental hope that ethics can transform our lives, so that most people, if not all people, can lead better lives. "Better" is of course problematic in this context: Morally better? Prudentially better? How does leading happier lives relate to leading lives in which we fulfill our moral duties? And how do we justify whatever answers are given these questions? Practitioners of ethics readily engage these questions. Among some, however, there is dissatisfaction with covering old ground, or fine-tuning responses to objections which themselves have been fine-tuned. This dissatisfaction echoes a similar dissatisfaction in epistemology. In the latter case, the analogous dissatisfaction is with the perceived fixation of epistemology on issues of skepticism. Inquiries as to the grounds of knowledge, the means of knowledge, the justification of knowledge claims are tasks which have largely occupied the discipline. Rather than pursue these tasks, some have turned away from such questions in pursuit of matters of "knowledge production," as evinced in enterprises such as naturalized epistemology and social epistemology.[1] Those engaged in these studies find it unsatisfying to worry the meaning of epistemic terms, or the justification of epistemic claims. They at times engage such concerns, but not as a main emphasis. Even as one might recommend that ethics get somewhere other than clarity and detail of distinctions and point-counterpoint of diminishing returns.

This essay is composed in the spirit of getting somewhere in ethics. The suggestion that ethics has not gotten somewhere is offensive and unintended. Appropriately put, the suggestion is that ethics should get somewhere further.

There are periods of intellectual activity in which clarification is paramount, and there are periods in which other sorts of breaking new ground is of primary concern. It is impossible to claim that humanity is in the latter period without seeming arrogant. So be it. For what additional period do philosophers want to debate the subtleties of utilitarianism versus deontology? Or to refine our understanding of normative terms with reference to common intuitions as the standard?

It will be responded that, even as epistemology has its nonstandard (by traditional standards) movements, so not everyone in ethics is engaged in what have been presented as the standard tasks. In fact, it seems likely that there has been more "nonstandard" activity in ethics than in epistemology. There is the overall field of applied ethics, comprising such subareas as biomedical ethics, business ethics, environmental ethics. There are discussions of feminism having ethical dimensions. Aspects of contemporary continental philosophy are deeply involved in ethical explorations. Taking account all of this, for what reason would one bemoan this field of philosophy, ethics, as nonprogressive?

The main response to this question will be given below. That "traditionalists" in English-speaking ethics tend to devalue thought under the rubrics of Applied Ethics and Contemporary Continental Philosophy is no news. Such attitude is to some extent dismissible as defensiveness based on commitment to a competing alternative. Dissatisfaction with "traditionalist" approaches has already been stated. From the viewpoint of wanting ethics to deliver an overall theory of a good life for everyone, or as close to a good life for as nearly everyone as possible, the alternatives of applied ethics and contemporary continental philosophy are lacking, although for opposed reasons.

Practitioners of applied ethics address practical questions, often as specific as individual cases. Applied ethics presents an image of problem addressing, if not problem solving. Its specificity is its strength, and its lack. Regarding most problems it addresses, their resolution does not constitute the fullness of a life. Granting that problems of medical treatment, professional dealings, environmental policy, etc., may have large bearing on individual and collective lives, their resolution, to whatever extent possible, leaves much unaddressed life as remainder. Further, there is not much indication or sketching regarding the living of this remainder. Satisfactory access to medical care, for example, alleviates some concerns, thereby enhancing one's life, but the majority of one's life often lies outside this concern.

Much recent continental thought aggressively addresses the overall problem of living one's life. Despite the frequent use of political vocabulary, however, there is at times an apparent self-centeredness to the discussion (consider, for example, the amount of focus on "the other" as a problematic "not-me"). Contrary to the continental philosophers' emphasis on subjectivity, one might maintain that the subjective/objective dichotomy admits of more likely reconciliation from an objective viewpoint, rather than a subjective one.

Such critical remarks are both stock and cavalier. Exceptions to these remarks can be pleaded (e.g., the political concerns of Sartre, Habermas and others of the Frankfurt School, Althusser). These exceptions are not without their point. Coupled with them are replies on behalf of practitioners of applied ethics, who might respond that their field is not intended to replace the more traditional body of ethical thought but rather to supplement that body in various respects.

These replies granted, one might remain dissatisfied with the results of ethical thought, taken in its entirety. The feeling that it is time to consider alternatives might lead to striking out in a novel direction. Or it might lead to a reexamination of previous work if the belief is that an important aspect of this work has been subsequently neglected. All this is heading to a proposed reexamination of Plato's goals in the *Republic*. There may be a manner of regarding these goals such that what Plato was attempting has not been continued or reattempted. It may further be profitable to compare Plato's work with the methods and accomplishments brought into being by the rise of modern science. The pursuit of this suggestion immediately follows.

At some time in the mid to late seventeenth century the systematized modern science of physics originated. It did not emerge from nothing, and it did not arrive in a final version. Although Isaac Newton is generally credited with its creation, it is likely that contemporaries of Newton contributed more to its inception than is commonly acknowledged. These qualifications and unsaid others granted, the accomplishment of Newtonian mechanics is remarkable in its transformation and redirection of much human intellectual endeavor termed "science."

Among the various ways to consider Newtonian mechanics, there is the following. The mechanics integrated various notions such as velocity, acceleration, mass, and force into a system of definitions and laws that allowed further elaboration, and its successes gave satisfaction to human desires for explanation, prediction, and manipulation. While much more can be said regarding this integration and satisfaction, this image of the success of Newtonian mechanics will serve for what follows.

There is reason to believe Plato may have been attempting a feat similar to Newton's, although in an apparently different area of thought. That is, Plato might have been seeking a system of integrated concepts of value, such that this system would satisfy desires similar to those satisfied by Newton's accomplishment.

The thesis that Plato is working along similar lines to Newton, expressed in this manner, seems obvious. Of course Plato is seeking integrated concepts of value for the betterment of humankind. Many, if not most, if not all normative philosophers are doing this. Still, with regard to normative thought there may be something special, if not unique, in Plato's endeavor. Recall the subtle questions raised in the beginning of the *Protagoras* (329c-e) regarding the logical character of the relationship of the excellences: whether they relate to one another somewhat in the manner of the parts of a face, or as the smaller golden parts of a lump of gold; whether they can exist independently of one another. Recall further the (re)definitions of notions such as courage (*Protagoras* 360), punishment (*Gorgias* 478), governing (*Republic* BkI), and justice (*Republic* BkIV). In the case of at least some of Plato's accounts of such notions, it is arguable that Plato is both aware and untroubled by what strikes the unfamiliar reader as obvious: that these accounts are not in keeping with common understandings of the terms. As with Newton, Plato's disregard of various common meanings seems done in the interest of delineating a type, or group of related types, for the pragmatic purpose of getting a handle on an area of experience. And as with Newton, Plato does not seek a complete departure from common notions. Even as one recognizes aspects of commonsense "force" in Newtonian "force," so Plato feels that the words of common thought reflect—albeit often dimly—genuine articulations.

One might yet believe that little has been said to distinguish Plato from a host of other normative theorists. Few would deny that there are distinctive elements to Plato's thought, but the claim that Plato is unique in suggesting integrated redefinitions of normative notions seems prima facie doubtful, if not absurd. Defending this claim is a lesser part of this chapter, in which the following theses are presented:

1. Plato does seek an integrated system of normative concepts.
2. In so doing, various normative concepts require redefinition.
3. Among recognized normative philosophers, Plato is virtually unique in doing 1 and 2.
4. Such work is not being done now (this seems a consequence of 3, except that it could be argued that although practically none of the major historical figures addressed both 1 and 2, these tasks are being addressed in contemporary discussions).
5. Such work should be done (most important).

The *Republic* is the work in which Plato goes furthest in presenting an organized system of normative notions. The (re)definition of courage made in the *Protagoras* is repeated, and various other normative notions are integrated. Regarding the project here ascribed to Plato, however, the *Republic* has an unfinished feel. Various normative notions are not considered, including several Plato has already labored. At best, in Plato's own words, the project appears hypothetical. Various notions such as justice and moderation are examined both as to what they are and how they interrelate (two tasks Plato likely does not want to separate). Yet despite initial claims to finality (*Republic* 444, but see 435 c–d), nothing is finally settled because the system as a whole is not given (511 c), neither as to all its parts, nor consequently their interrelation, nor is it shown that this system is the best (the idea of the best, The Good, is itself an hypothesis not realized in the *Republic* 505 a).

After the *Republic* there is no further work on this project. In the *Sophist* Plato integrates (or "interweaves") the notions of being, sameness, and difference, but he does not alter their definitions (except insofar as the interweaving of them has definitional consequences). In various later dialogues Plato displays the method of division, which together with the overall collection of the division is integrative, and hence of definitional import. But there is no grand unification of the sort suggested in the *Republic*, and no revolutionary redefining of the notions under examination. It is hence possible that Plato did not intend the grandiose scheme ascribed to him here, or that having intended it he later abandoned the project. Alternatively, Plato might have seen the enormity of the project and bequeathed its fulfillment to future generations, instead pursuing other, at times related, matters. These choices can be debated, perhaps largely settled, but such will not be attempted here.

Whether the view of Plato's project presented here is accurate or mistaken, this view of Plato is the inspiration for what follows. However, the correctness of this reading of Plato is not of final importance. It will be suggested that what is of importance is that this vision, ascribed to Plato, be continued.

The intention to pursue what has been presented in Plato's name is met by two conflicting objections. Hasn't the project of redefining and integrating ethical terms so as to form a coherent system of ethics been the agenda of many, if not most, moral philosophies? On the other hand, is this project worthy of pursuit? Would it not be philosophically preferable to uncover and refine common meanings, rather than seeking redefinitions which at best will result in a coherent system having strained relation to our ethical intuitions?

Universal claims regarding philosophical practice are difficult to establish, and often easy to controvert. To support the claim that no one of note has taken up the task ascribed above to Plato, consider the following. Many ethical theorists have concentrated on what seemed to them a single key term, including at times several terms related to this key term. This is similar to what would have been the case had Newton concentrated on the term "force" without reconceptualizing the various interrelated metric and kinematic notions which together constitute Newtonian mechanics. The monumental advances such as Newton's theory of universal gravitation occur within the framework of Newtonian mechanics, and could not occur without some such framework. If the analogy is pertinent, then attempts at clarifying comparatively singular notions such as "right," "good," "morality," and "justice" suffer the supposed fate of the counterfactual Newtonian fixation on the notion of force, without the necessarily related systematic revision.

So, for example, following others, Mill proposes an ethics based on pleasure. His proposal is systematized to the extent that the good is what we all desire, and our duty is in some respect to promote the good. Further, justice is considered in relation to the good. In all this, Mill believes he is clarifying and systematizing our ordinary intuitions. However, there are a host of notions related to living well about which Mill says nothing. Courage, for example, is not discussed in *Utilitarianism*, nor is temperance (except to the extent that cultivated pleasures are favored over crude pleasures).

Initially, Kant's moral theory appears to be what is sought here. It is systematic, and to some extent it involves redefinition of terms such as moral righteousness, insofar as it had not occurred to previous theorists to connect the notion of moral rightness to natural law in the manner in which Kant connects these notions. Yet to a greater extent than Mill's theory, Kant's is noncomprehensive. Kant was aware of this lack, and considered it a positive feature of his theory. Kant believed that to worry terms such as happiness, pleasure, delight and amusement is to be caught in a "heteronomic" space of shifting desires and opinions. Hence, in this respect Kant's theory may be regarded as antithetical to what is here postulated as Plato's.

Aristotle does seem to be attempting a comprehensive ethical theory. However, Aristotle settles for more or less common understandings of ethical notions. So, while he does integrate many notions and is systematic, the result is essentially a refinement of common thinking. It is as if Newton somehow systematized commonsense notions of force, mass, velocity, acceleration, and so forth.

Of the traditional philosophers Spinoza perhaps most closely approximates the undertaking ascribed to Plato. There is systematic integration of technically defined notions in the *Ethics*. There are also a set of core ethical terms, among which are "striving," "passions," and "reason." In Book III of the *Ethics* he presents a fairly comprehensive proto folk-psychology. Clearly, Spinoza ties his ethical theory to his view of human nature, such nature being continuous with nature in general. All this is to Spinoza's credit, in the spirit of this study. There are nevertheless the following two significant differences between Spinoza and Plato. First, however confidently the character Socrates presents Plato's views, there is the occasional admonition that the systematic theory to which Socrates alludes in the middle books of the *Republic* (not even sketched, simply indicated) is a grand hypothesis realizable—if at all—only under the right future conditions. This difference places Spinoza nearer to Newton than Plato in terms of the confidence with which the scheme is delivered. As will be argued below, however, Spinoza's confidence is misplaced. Second, Spinoza does not seek a theory of universal betterment. Much of Spinoza's work is descriptive-explanatory. Spinoza does not think many will follow the path to what he terms "freedom," and he offers scant aid to those who remain in the grip of passion. Granted that Plato vacillates on showing concern for the common folk, his stated purpose is to transform society so as to maximize human well-being. Were material conditions such that labor-intensive work could largely be mechanized (unimaginable in Plato's time), one suspects that Plato would attempt to incorporate the many into the best sort of life.

There are too many moral philosophers to examine here, even in the cursory manner in which Mill, Kant, Aristotle, and Spinoza were surveyed. This is so even if the search is restricted to pre–twentieth-century Western moral philosophers. Adding Eastern philosophers, as well as contemporary and near-contemporary thinkers, prompts the supposition that someone satisfies the terms of the task ascribed above to Plato. Or, if not some one thinker, perhaps a combination of philosophers. Considering the emphasis on the essay format prevalent in recent professional philosophy, it seems plausible that an understood division of labor allows individuals to work on various aspects of an overall task, as is the case in the natural sciences. Of course, this last suggestion supposes an understood common task, beyond the comparatively wide guidelines for publication in professional philosophy journals.

Nonetheless, there is substantial reason to believe that Plato's vision has not been realized. Succinctly put, there is as of yet no heaven on earth, nor does such appear a future possibility towards which we are converging. Put another way, humanity has yet to embrace a successful ethics. Since these statements will strike many as either ludicrous or outrageous, elaboration and defense are in order. "Heaven on earth" intends the state of affairs in which a significant majority of the world's population lead self-satisfied lives at their maximum potential. Such potential is defined in terms of whatever constitutes the current state of the art of human practical and theoretical knowledge. Such potential involves an awareness of this current state of the art. So, for example, a significant population of self-satisfied people living in backwater ignorance of current human achievements would not be a state of affairs of heaven on earth. People living in resentful poverty would not qualify for this state, nor would those living affluent lives, separated from and fearful of others perceived as resenting them.

If the majority of the world's population is to lead knowledgeable lives, a surplus of material wealth is necessary to allow time for such acquisition of knowledge. Such a surplus creates problems of distribution, including the socio-psychological problems of envy and resentment. Such a surplus also invites environmental degradation through the productive processes realizing the surplus. If such problems are unresolvable in principle or in practice, visions of heaven on earth must be abandoned.

Problems of human nature have been alluded to in the form of envy and resentment fostered by distribution of wealth. There also arise questions regarding the egalitarian form of this vision of heaven on earth. Problems, if not outright paradoxes, of choices among distribution schemes are well known. Various thinkers, perhaps preeminently Nietzsche, challenge the assumption that the majority of humanity can lead exemplary lives. Nietzsche further challenges the assumption that self-satisfaction is an ingredient of the best life. Thinkers as diverse as Sartre and Schlick challenge the putative factual reference of normative terms.

This quick survey of problems appears tactically designed to lessen criticism by means of brevity. One might insist that the above-mentioned difficulties, incomplete though the listing be, if adequately developed would force abandonment of heaven-on-earth talk. Countering these developed criticisms would be a daunting task. Further, given that it is part of the presented thesis that no scheme for heaven on earth has yet been discovered, these criticisms can only be addressed in terms of the possibility or probability that they can be overcome.

To address the criticisms in such terms, consider once again the analogy between the hoped-for inception of a successful ethics and the rise of modern science. Many, if not most, historians and philosophers of science would deny the proposition that genuine scientific activity begins with modern science. Examples of ostensible scientific activity predating modern science include Ptolemaic astronomy, Aristotelian physics and biology, and some aspects of alchemy. Arguing the legitimate range of the term "science" would greatly extend this discussion. Let it be granted, instead, that the above examples are instances of genuine scientific activity, in which case such activity precedes the rise of modern science. These and other such examples show the limitations of such sciences, when compared to their modern counterparts. A largely planetary astronomy—regardless of whether its orbit functions are well chosen—cannot match the generality of an astronomy not so limited to the planets and so forth for the other examples.

It is undeniable that there are accomplishments in ethical theory, even as it is undeniable that there are accomplishments in earlier natural science. There is some limited consensus on key notions, as there was in the pre-Newtonian period of natural science. Although, as in that period of science, there is vigorous debate regarding the characterizations of these notions. The question arises as to how Newtonian physics supplanted competing scientific practices to such a large extent within their domain of concern.

It will likely be noticed that the above discussion is reminiscent of Thomas Kuhn's comparison of scientific progress with the perceived lack of progress in philosophy:

> If we doubt, as many do, that nonscientific fields make progress, that cannot be because individual schools make none. Rather, it must be because there are always competing schools, each of which constantly questions the foundations of the others.[2]

In citing Kuhn, no endorsement is implied about Kuhn's antirealism concerning scientific truth, or his suggestion that other intellectual endeavors, such as philosophy, will always evince fundamental conflicts. Yet even those in strong disagreement with Kuhn can acknowledge the usefulness of his discussion of scientific accomplishments. In particular, Kuhn struggles with the question as to why a certain outlook comes to be "paradigmatic" of scientific practice within its field. What is noteworthy here is the generality of his answer. Certain critical problems seem best approached in the new manner. There is promise of further results, and much work remaining. Kuhn avoids saying the new research program gives insight into the truth of things, but consistent with his inclinations he could say, although he does not, that the scientific community adopting the new outlook believes that such insight has been given. Such a drive for truth, along with other standard reasons such as desire for description, explanation, prediction, and control, go some way towards accounting for the adoption of an outlook.

There is nothing in Kuhn and related thinkers to suggest that adoption of a new outlook is in any sense mechanical. To that extent, such thinkers present a united front against the logical positivists' dream of a rational reconstruction of scientific practice. Whatever one's realistic inclinations, Kuhn reminds us of what it was like on the other side of the important discoveries, the side prior to the discovery. One lack of Kuhn's account, however, is that it does not often emphasize the relative importance of some discoveries (institution of a "paradigm"), vis-a-vis other discoveries. The success of Newton's quantitative procedures, for example, when taken together with prevalent corpuscularism, likely aided the rise of Daltonian chemistry. This is to say that there are key discoveries which set the intellectual climate within their general domain, at times for succeeding centuries.

Within the past several hundred years there have been strains of ethical thought, broadly construed, that have had powerful effects on human affairs.

Classical liberal thought and Marxism, both considered here with regard to their ethical content, are among the examples of such influential thought. Yet none of these intellectual movements can be said to dominate the intellectual landscape of ethics to the extent of largely ruling out competing thought. It seems as though ethical thought will not duplicate the accomplishments of natural scientific thought until it becomes itself scientific.

Opposition to this last suggestion is swift and sure. The epithet of "scientism" seems applicable. Further, if Kuhn's thought is appropriated in support of this last suggestion, it should be recalled that Kuhn does not deem natural scientific theory sacrosanct.

To some extent this criticism is appropriately aimed, whether successful or not. The tenor of the discussion has been that ethical thought ought to duplicate the success of natural scientific thought, and that it has not yet done so. If the suppressed premise be added that only scientific thought can succeed in the manner of scientific thought, then the charge of scientism is not groundless. On the other hand, the success of natural scientific thought has not banished philosophical reflection on natural science, as Kuhn and others have shown. Even if ethics becomes a largely scientific discipline, there will be much upon which to philosophize. The very claim that ethical thought *ought to* duplicate the success of natural scientific thought, for example, invites philosophical examination. As for Kuhn's "desanctification" of science, he has not shown science any less successful in its pursuits of explanation, prediction, and control—nor has he intended such. If he is accepted on this point, Kuhn has only shown science not to be productive of final truths.

Why suppose that ethics can be successful in the manner of science, thereby profoundly transforming human life in a universal or near-universal manner? To some comparatively limited extent, as noted above, ethical thought has been transformative. What is required is the universal, or near-universal, perception that the delivered successful ethical theory has got matters right. Any argument in support of the likelihood, or even possibility, of such acceptance will be psychological and/or sociohistorical in flavor. Such an argument will hence lack the logical strength of an a priori justification. Students of Plato's *Republic* might recall here the brevity of Plato's sketch of the implementation of his grand scheme (540d–541), indicative of his worries in this matter.

The very possibility of such an argument is challengeable. Modern natural science, most notably physics, arguably took hold because of ruling class interest in ballistics. In general, science was a gentleman's pursuit, and most of the European population were not gentlemen. It is easy to claim that there are no similar circumstances regarding ethics presently or in the foreseeable future. Now it is nations and supranational industrial and financial interests that seek their forms of dominance. These forms of dominance have room for science, but little for ethics.

Sweeping claims such as that the rise of modern science is a result of ruling class interests are difficult to assess. But even if other factors were key in the development of science, it remains incontestable that the generally illiterate seventeenth- and eighteenth-century population of Europe had little information or interest regarding scientific developments. Until recently, the mode of life for most populations, even in comparatively developed societies, has been one of immediate concerns and performing daily, repetitive tasks. Education and leisure time have come late in the day, and with them has come exposure to scientific accomplishments, leading to socially influenced acceptance of prevailing scientific outlooks. Especially as science has been mediated by engineering technology, and resulted in medical advances, populations have generally viewed science as in their interest, and accepted it accordingly.

There is similar promise of widespread acceptance of a successful ethics. Nowadays one not only speaks of worldwide literacy, but growing worldwide computer literacy. Information is arguably managed by and for the propertied class, but communication across worldwide computer networks has perhaps for the first time realized Mill's ideal of a marketplace of ideas. One can no longer expect restricted access to ideas of general interest. There is fertile ground for promulgation of a successful ethics, even in the face of conflicting, short-sighted ruling class interests.

Access to ideas is of course a necessary and not a sufficient condition for the universal acceptance of what will be a successful ethics. Whether developed piecemeal or all-at-once, there must exist such an ethical theory to be disseminated. It has already been stated that no such theory has been publicized, although Plato has presented the general idea. Hence, the form of such a theory, if such a theory is possible, can only be speculated. What follows is such speculation.

A successful ethics will be grounded in an accurate and appropriately comprehensive theory of human nature. "Appropriately" comprehensive because there will likely be various matters, biological for example, that are not of immediate concern for such a theory. As Plato saw, such a theory will necessitate an adequate psychology. While Plato did what he could to supply a psychological theory, it is arguably the case that such a theory has not yet been developed, nor is there agreement among practitioners as to the manner of such a theory (neurophysiological? behavioral? other?). The manner in which a theory of human nature would support a successful ethics has been challenged by Sartre and similarly-minded thinkers. If their claim is correct that no such theory of human nature is possible, then a successful ethics as conceived here is not possible.

To proceed, it is assumed a psychological theory adequate for the task of conceptualizing human nature is possible. Whether the theory will be folk-psychological, "classically" computational, neurophysiological, neurocomputational, psychoanalytic, phenomenological, some combination of these, or something else—perhaps hitherto unknown—cannot be decided here. Again following Plato, it is likely that a major component of a successful ethics will be sociopolitical. This is so as long as human nature, however it be scientifically accounted, remains as it is. If human nature is altered, there will remain a sociopolitical component of a successful ethics as long as the human situation is one of interdependence.

A successful ethics might necessitate alterations in human nature, either wholesale on the entire human population or on an individual basis as needed. Alterations in human behavior are routinely accomplished by social means such as education, and in individual cases by means such as pharmaceuticals. Humanity is likely on the verge of important breakthroughs in understanding brain architecture, neural message coding, and related matters. With such understanding comes the possibility of further alteration and control of human behavior. Should such possibility enter into the discussion of a successful ethics? There are arguments on both sides of this question. Certain notorious utopian schemes are often cited as warnings against moving in the direction of behavior modification.

While we cannot ignore judgments on previously offered utopian schemes, we ought to inquire into possibilities increased knowledge lends. Suppose we come to have strong reason to believe that some of the fundamental problems of human life are insoluble, given present human nature. This contentious supposition includes such assumptions as that what constitutes a problem of human life is agreed upon in the manner in which propositions of entrenched scientific theory are agreed upon; that some form of hierarchical ordering of such problems is similarly agreed upon; that what constitutes human nature is similarly agreed upon; and that prevailing human nature necessitates such problems are similarly agreed upon. If all this can be accepted, suppose further that we possess the means of alleviation of such problems through alteration of human nature, while simultaneously retaining, and perhaps enhancing, agreed-upon positive aspects of human life.

As an extremely limited example of what is being proposed, consider the phenomenon of road rage. People who often are otherwise normal seeming lose their humanity in stressful driving situations. Despite appearances to the contrary, such behavior is likely not an isolated event in an otherwise serene psyche. Before we state with confidence that the world would be better without such "hot-tempered" characteristics sprinkled through the population, we need think in a somewhat ecological manner of the place of such characteristics in society, as well as in the individuals possessing them. If careful thought, making use of concepts and laws that for the most part have not yet been formulated, were to determine an affirmative answer to the question of the desirability of tempering such natures—both for society and the individual—then if means for mitigating these characteristics were available, it seems they ought be availed.

Returning to Plato, recall that his rigid and likely unworkable three-tiered society in the *Republic* was based upon his view of the distribution of human talents. Such a view might have been encouraged by his understanding of the then-current possibilities of production and distribution. As previously noted, Plato might have considered it necessary for the development of a comparatively few philosopher-rulers that the majority of the population be laboring to produce a sustainable surplus for the support of such an intellectual elite.

Would Plato alter his scheme if he were aware of current and foreseeable future possibilities of production? He repeatedly states in the *Republic* that one must not dogmatize about structural and procedural matters (for example, 416b, 473a). If by current lights the rigid class structure of his proposed society seems contrary to the best that can be given to the most people, it is easy to suppose that a twenty-first-century version of the *Republic* would radically depart from the original. What would be kept in all likelihood are the suppositions that governing and social organization are matters of science, and that practitioners must be trained in this science, or sciences, for the benefit of all.

The scheme of the *Republic* is based upon the supposition that governing and living are matters of knowledge. Plato gives the impression that such comprehensive knowledge is at hand, or near at hand. We know now that in Plato's time it was not. If the thoughts expressed here are correct, we are still not in possession of this knowledge. We must of course carry on, relying on our intuitions of what is proper and best. Some matters appear settled, even as some matters in alchemy and Ptolemaic astronomy carried over to modern chemistry and astronomy respectively. But the general scheme is wanting. Breakthroughs in the behavioral sciences must be coupled with bold initiatives in value theory, if life on Earth is to be significantly improved.

NOTES

1. A vigorous statement of this philosophical attitude can be found in Steve Fuller, *Social Epistemology* (Bloomington: Indiana University Press, 1988), Preface.

2. Thomas Kuhn, *The Structure of Scientific Revolutions*, 2d ed. (Chicago: University of Chicago Press, 1970), 162–63.

Chapter Three

On Human Improvement

> In order to cure most of the ills of human life . . . Let the whole species possess naturally an equal diligence with that which many individuals are able to attain by habit and reflection . . . —David Hume, *Dialogues*

The thought Hume's character Philo expresses regards the possibility of human improvement. Many people think that most individual humans are capable of improvement. Of course, people are not always able to specify clearly their intuitions regarding such improvement. And Philo is not primarily thinking about improvement on an individual-by-individual basis, but rather an overall improvement in selected aspects of human nature, which nature overall might be said to constitute the human species (or, if preferred, to be the class-defining characteristic of the species). In making this suggestion, Hume's Philo is not alone. Philosophers as diverse as Plato and Nietzsche have spoken of such general human improvement. Religious prophets have demanded it.

It is reasonable to suppose, however, that in anything but the most general terms there is no common agreement about what constitutes such longed-for human improvement. Nor do I intend to unify the discussion here. I do intend to advance the discussion by offering speculative suggestions about a kind of human improvement, and surveying the critical environment of these suggestions. Primarily, my aim is to start a discussion, rather than offer a polished version of a solution to an age-old problem. For those who find the discussion frivolous, I shall offer some comments towards the conclusion of this chapter.

In brief, I want to examine the claim that people can be improved through internal, most likely physiological, modifications which enhance their cognitive abilities. To a lesser extent, I want to examine the claim that similar modifications could enhance a person's moral capacities.

That the human situation is capable of improvement, and that it ought be improved, are not controversial claims. Controversy starts at the first step of specification. Are we to evolve towards some state-of-affairs we have not yet instantiated, or towards some circumstances we have largely occupied and departed? And whichever direction, what is the relation of this discussion of proposed heading (an example of such: further large-scale industrialization, or movement towards de-industrialization?) to the question of the improvement of human nature?

This last question is raised to indicate the enormous domain of the topic of human improvement. Let us constrain the topic somewhat by focusing on improvement in human nature, rather than improvement in external circumstances surrounding human life (but see below). So to delineate the question is not to deny the influence of external circumstances, including social circumstances, on human nature, but rather to speak of other sorts of causes and effects. So limited, the domain is still enormous. To speak of improvement of human nature is to speak normatively, and to speak normatively about human nature is to enter a large arena in which the task of specification of human nature might be considered non-normative (descriptive), whereas any case for improvement of human nature would likely be considered normative (prescriptive). We then confront the question of reconciliation of such factual and normative concerns. In brief, we face an ethical inquiry regarding a topic of such scale as to involve many of the outstanding questions of moral philosophy. This enormous task will not be attempted here. Rather, the suggestion will be made that such a task is largely avoidable in the context of various focused discussions regarding the improvement of human nature.

The assumption that the specification of human nature is a non-normative undertaking has been questioned, perhaps most notably by Sartre, who maintains that we are ontologically free to define human nature by our choices. To an extent Sartre is correct; it will be urged here that human psychophysical nature can be viewed as partially a matter of choice. But this choice occurs within natural constraints, and to this extent Sartre is mistaken (it may be argued that Sartre recognizes the difference between natural constraints and the range of human choice, but at times he seems to ignore the extent to which natural constraints are inseparable from what is intuitively recognized as human nature).

Since I shall propose improvements in human psychophysical nature of the sort that enhance cognitive powers, I cannot wholly avoid the criticism that such suggestions are misplaced. It will be maintained by many that to improve humankind internally, to make better people—cognitively, morally, affectively—one must primarily address the conditions in which they live. To assume blithely that cognitive human improvement can be posited as a goal to be studied independently of supposed environmental causes (presumably on the grounds that "synthetic" means of achieving these effects will be found) is to commit the following two related errors: (1) assuming that however complicated the effect, a radically different sort of cause can be substituted; and (2) assuming that specifics of the environment are completely separable from the internal characteristics of human life. These two assumptions are felt erroneous to the extent they require "brain-in-a-vat" sort of technology (i.e., regardless of the actual environment, the brain can be manipulated so that it is mindful of any desired environment), and to the extent they assume that the implementation of such technology would be an improvement over current circumstances.

It might be felt that the improvements I shall suggest do not differ from the apparently undesirable "brain-in-a-vat" situation. Perhaps in the distant, unforeseeable future, it comes to pass that something like "brain-in-a-vat" technology is employed to improve human life. However, concerning the foreseeable future, the use of such fanciful technology is not currently regarded as generally improving human life. On the other hand, there is no reason to suppose that needed improvements in the external circumstances of human life cannot be conjoined to enhancements in human cognitive powers achieved by means yet unknown or untried. Clearly, one may suppose the possibility of such internal enhancements while prioritizing the improvement of external circumstances. We all suppose that improvement of external circumstances (e.g., nutrition, housing, social services), at least above various threshold levels, generally tends to enhance human cognitive abilities. But we might break new ground by asking, "Is this all? Are there ways of improving human cognitive abilities which are more internally directed?"

Improving a person's cognitive abilities does not obviously necessitate the improvement of their moral nature. The relation between cognitive improvement and moral improvement is sufficiently complicated that much remains to be known, and whatever is considered must be carefully qualified. There is the further question regarding the relation of moral improvement to general improvement of quality of life, but to simplify the discussion somewhat let us restrict ethical concerns to the question of moral improvement. Of course, people disagree at various points as to what constitutes moral improvement. These observations lead to the following general criticisms of the implicit suggestion that the improvement of human cognitive abilities thereby improves human life. First, if the nature of moral improvement is not settled, how can we be certain that internal improvement of our cognitive nature will tend to make us morally better, rather than morally worse individuals? Perhaps there is a limit to how "intelligent" we can become, on average, and still be the general sort of moral beings we are. Second, our inability to specify the nature of moral improvement to an exact and final degree suggests that all sorts of human improvement, including cognitive improvement, are similarly unspecifiable.

To the first criticism, it must be conceded that internal changes in our cognitive abilities will have effects on other aspects of our nature which are to varying extents unpredictable. It may seem initially plausible that, roughly put, the smarter we are the morally better we are likely to be, all else being equal. However, the ceteris clause is in question because speculated cognitive improvements, such as will be considered below, are unlikely to leave all else equal. And it must be admitted that we don't know for certain the consequences of general improvements in human cognitive abilities. To speculate by rough example, suppose IQ tests are a reliable indicator of broad-spectrum cognitive abilities. Suppose now that everyone's cognitive abilities can be brought up to current "genius" IQ standards. Can we be certain that in such circumstances people will be more generous, fair-minded, caring, rather than more miserly, cunningly unjust, and self-centered? Given present knowledge, including ethical as well as sociological/psychological theory, it is but a hope that powerful improvements in general cognitive abilities will promote what most would consider to be overall improvement in moral behavior.

We can at least postulate the following. If Plato was correct in thinking that morality is a function of knowledge, and complex knowledge at that, then people with improved cognitive abilities are at least in this respect better positioned to lead moral lives.

Regarding the second criticism, there are specific examples of human improvement where the goal, if not the means to the goal, are clear. Typically, such goals concern physical performance. There is little dispute as to the meaning of jumping higher or longer, for example. Various other skills have a degree of convention in their specification. For example, speed of keyboard typing must be weighed against accuracy of entry. Such complications as degree of arbitrariness in weighting potentially conflicting criteria open the discussion to criticism, but if such conventional choice is subject to some constraint (for example, no one will choose typing speed at the expense of 100 percent inaccuracy), then we can say that the conventionally stated goal has some intuitive plausibility, and is in any case capable of clear, quantitative specification (at least in the typing example). Can it not be urged that aspects of cognitive performance have similarly clear goals, and intuitive plausibility regarding the range of acceptable settlements in cases of conflicting criteria? Consider, for example, speed of thought. Restricted to various domains, such as mathematical problem-solving, faster thought is intuitively preferable to slower thought, provided comparable accuracy.

The claim that widespread prima facie improvements in cognition, such as faster computational thought, are bona fide general improvements in the human situation, is controversial. It might be urged that such faster thought is in potential conflict with more profound thought. To speculate, consider an eminent physicist such as Einstein, who was not the most accomplished mathematician, and reportedly not the fleetest at calculation. Would we desire a physicist of Einstein's profundity (allowing for the discussion a loose term such as "profundity") but fleeter of calculation? There is the possibility that such fleetness, even as an unactualized potential on any given occasion, might to some extent conflict with profundity (and so a fleeter Einstein might not have been the one to discover general relativity). The question as to the relation of fast thinking to profound thinking is highly speculative, and charged with vagueness (given the vagueness of "profundity," as well as the need to further specify "fleetness at calculation"). But perhaps the vagueness can be resolved, although such resolution might produce a spread of clear alternatives, rather than one clear question. There remains the question as to whose question it is, or similarly, what sort of question is it? Is this a question for psychologists? Suppose there is a concealed conceptual incompatibility between such modes of thought. If so, perhaps the question is one to which philosophers may lay some claim.

To appreciate the possibility of philosophical content in such questions, consider the topic of conscious parallel thinking. This topic has some relation to that of fleetness of thinking, inasmuch as parallel processing is, in some respects, faster than serial processing. While there is little doubt that unconscious systems such as "early" visual processing are parallel in the sense that a multitude of neurons are simultaneously transmitting to the brain, there is less certainty that conscious thought instances parallelism at the level of consciousness. Of course, to speak of consciousness is already to enter the philosophical lists. Allowing for the moment the unquestioned existence of philosophically full-blooded consciousness, the parallelism of such consciousness within an individual invokes philosophical speculation. Let it be assumed that, were it possible, concurrent conscious thought comprised of two or more separate threads of thinking would, in general, enhance the speed of thought. What we are considering is a distribution of conscious intellectual labor within the subject, such that the subject divides various problems to be simultaneously addressed in parallel conscious tracks, and then integrated as needed. For example, the hypothetical parallel thinker might simultaneously be doing taxes, planning dinner, and composing a short story in which the actually planned dinner will appear fictitiously.

To note one set of philosophical problems from among the many raised by the above speculation on parallel conscious thought, consider the possibility of a parallel-conscious self.[1] To the extent that a serially unified consciousness is a necessary condition for selfhood, it follows that a parallel-conscious self is no self, or not one self. Of course, much has been written regarding self-identity, and nothing can be said here that is uncontroversial. Let us simply note that it is not clear that parallel consciousness in a single human-brained body entails the improvement of a single human self. There may also be deep phenomenological reasons why a parallel consciousness is logically impossible. Such reasons may have less to do with the concept of an individual person, than with the nature of consciousness itself.

This brief look at the philosophical problematic attending discussions of parallel consciousness indicates that suggestions of radical alterations in human cognitive processes ought to receive philosophical scrutiny. Some of the philosophical problematic has to do with the contexts in which an ostensibly cognitively improved individual (or set of individuals) are considered. Improvements in a sprinter's speed are unquestionably welcome, all else (such as length of career) held constant. Similar improvements in running speed might be more a mixed blessing in other sports involving running, given considerations of team coordination, among other concerns. If we look to speed up cognitive processes we should expect to encounter analogous difficulties of coordination.

We are approaching the time when science and technology will present us with choices regarding cognitive improvement (we already have various choices regarding noncognitive physical improvement, and we can expect more dramatic choices of such physical improvement in the future). We may choose not to avail ourselves of such ostensible improvements in our cognitive apparatus. However, we currently make the choice of assuring adequate protein intake for infants and supplying a stimulating environment for them, so as to bring their cognitive abilities to a certain normal level, roughly speaking. Assume the existence of procedures to inhibit cognitive abilities (lobotomy, for an extreme example). No one can be certain that the survival of humanity would not be better served by some degree of dulling (assuming we can agree on the intuitive notion of "dulling") of our cognitive abilities. We might become, say, more affable, more courteous, while retaining enough of our scientific and technological abilities to survive at some level that, from our current perspective, most of us would consider acceptable. This reflection shows that, at best, this discussion is probabilistic, rather than demonstrative. Constrained as we are to choose on a less than certain basis, do we choose enhanced cognitive abilities, neither enhanced nor diminished cognitive abilities, or depressed cognitive abilities? Humans have rarely chosen less of anything thought positive, and it is assumed that enhanced cognitive abilities are thought positive. Such an observation does not justify the choice for enhanced cognitive abilities in any absolute moral sense of justification, nor does it assure that, were significantly more information available as to consequences, the choice of cognitive enhancement would appear the best choice. But perhaps given our epistemic circumstances, it is the only justification obtainable.

Having reviewed some difficulties attending the notion of general improvement of human cognitive abilities, let us turn to positive reasons for pursuing such improvement. A stirring vision of widespread human improvement is enunciated by Trotsky in *Literature and Revolution*:

> Man will become immeasurably stronger, wiser and subtler; . . . The average human type will rise to the heights of an Aristotle, a Goethe, or a Marx. And above this ridge new peaks will rise.[2]

Trotsky believed such improvement would be implemented after the envisioned worldwide socialist revolution. More will be said below on the idea of improvement by means of external, social change. For now let it be said that we may share Trotsky's vision, while disagreeing about sufficient conditions of implementation.

We believe in the goal of universal literacy. Few would claim that literacy is beyond the reach of most individuals due to inherent cognitive limitations. We expect preadolescent children in industrialized societies to be literate. Consider the expectation that children before adolescence be literate in science to the extent that organic chemistry, relativity and quantum theory, molecular biology, and so forth are understood by them to such degree that they can routinely solve the sort of exercise problems currently given in texts read by graduate students specializing in these various disciplines. That is, the average child manages this for all these disciplines, as the average child today manages adequate performance in arithmetic, basic reading skills, basic knowledge of national history, geography, and so forth. While there have been rare cases of child prodigies, few would consider such speculated ability within the range of most adults, to say nothing of children. Perhaps such generalized ability is currently possible, given adequate motivation and a revolutionized climate of learning. Let us suppose that such ability is not currently possible, for if we are mistaken in this supposition we can easily imagine more impressive cognitive abilities than these, until a point of no reasonable dispute. If we consider it beneficial that children have such ability, and that adults have even greater ability, then we ought be willing to consider as yet wholly untried means, provided there is some indication of feasibility.

Imagine a world of such universal scientific literacy. Imagine similar acquaintance with humanistic texts, artifacts, and ideas. Imagine heightened abilities to interconnect, to hypothesize, to create new concepts, to celebrate aesthetically these powers and what they discover. Imagine these powers brought to bear on the concern that humanity live peacefully and cooperatively. This concern is perhaps the ultimate normative goal of social science. Imagine genuine progress made towards this goal, such progress being guided by what are perhaps as yet unknown psychological and sociological theories of such matters, or by some other science that has been found to do the work of directing us towards the aforementioned goal. Imagine, in short, a better world in which we have learned consistently and continuously to get on with one another and with our individual selves, even as we heighten our understanding of nature and our ability to manipulate it for purposes that either aid, or do not conflict with, such living and understanding.

This vision verges on utopianism and is subject to the general criticism of such visions; namely, that it is unrealizable. However some utopian visions may be approximated to varying extents, and this vision might admit of such approximation. The controlling assumption is the Platonic belief that living well is partially a function of knowledge, together with the good fortune of living in a society that supports living well, such a society also being a function of knowledge. Plato may or may not have despaired of establishing, or even approximating, an idealized society. Whatever his true belief on this matter, his discussions of human nature suggest that he felt constrained to work within the confines of such human nature as can be optimized by appropriate social mediation. He did not speculate on alterations that would allow for a society of individuals with basic cognitive powers far in excess of a Plato or a Theaetetus, nor could he be expected to entertain this possibility.

Perhaps this utopian vision is excessive. We still should face the possibility of a general heightening of inherent human cognitive abilities, if it becomes likely that this heightening can be accomplished. Suppose no grand vision is present, but as individuals we have the opportunity to enhance our ability to think. Experience suggests that the exercise of such choice is to some extent a function of the effort in the exercise. There are any number of subjects about which our ability to think will be improved through our commitment to a standard academic regimen focused on the subject. Many people are aware of this avenue of improvement, desire such improvement, yet are unwilling or unable to invest the requisite time and effort. Alternatively, most people will correct standard eyesight problems in standard ways, and as radical means of improvement (e.g., laser surgery) become more available in terms of cost and minimized risk, more people are expected to choose them. People now choose various pharmaceuticals on the basis of ill-established claims of memory enhancement. It is probable that if these claims were established, and risks shown to be minimal, a significant number of people would ingest these chemicals. In short, in the absence of any utopian expectations about society, people indicate a willingness to improve aspects of their cognition provided the expenditure in terms of cost, risk, time, and effort are kept at intuitively low levels.

There is ample ground between grand social results and pride-of-ownership personal improvement. In the *Protagoras*, Socrates attempts to demonstrate to the sophist that, on Protagoras' hedonistic assumptions, a nonrelativistic knowledge (measurement) of pain/pleasure is necessary for successful pursuit of one's advantage (356a–357c). Similarly, it might be held that one enhances the probability of a modicum of success in life, either in one's prudential or moral choices, if one chooses on the basis of the best available information. It is often noted, however, that there is a near-infinite amount of information connected to many of life's decisions. One might question the appropriateness of heightened rational deliberation in making real choices. In the related discussion of "look ahead" strategy with regard to game playing computational machinery, it is sometimes urged that a more human sort of play, using an intuitively restricted look ahead search for appropriate game choices, would be preferable to brute-force decision-tree pruning. Since real life is arguably more complex than games such as chess, all the more reason why enhanced cognitive abilities are misdirected, if employed in standard rational decision making procedures.

It would likely be easier to make the case for enhanced cognitive abilities if the appropriate method for making life's choices clearly were by means of rational decision making procedures. And it has not been settled here that such procedures are typically not the most conducive to success in life's choices, however success is considered. But whatever the truth of this matter, if decision making is generally best performed in a less computational manner, it still seems probable that such decision making will be improved if cognitive capacities are enhanced. The basis for this claim is that, ordinarily speaking, decisions are either made largely from learned habit, or they involve processes generalized by the term "deliberation"—and they may of course involve both. We are admittedly begging important questions in relying on folk-psychological concepts such as decision, habit, and deliberation, but provisionally proceeding on this basis, the following may be claimed: decisions made by deliberation are likely to be better if made with enhanced cognitive capacities. A more natural deliberation than that of strict rational computation yet may involve memory retrieval, interconnections of items, some amount of look-ahead, natural deduction, some degree of factor analysis, and so forth. Decisions made by habit are at times, in a cool hour, profitably subject to review by deliberation, and habits may be altered to advantage in this manner.

The notion of enhanced cognitive abilities can be understood through reference to cognitive psychology texts, such as Ulric Neisser's influential text.

Neisser discusses pattern recognition, focal attention, visual memory, speech perception, and various cognitive activities, such as problem solving, which draw upon the former abilities. Given the rough, experimentally-determined means and limits Neisser cites with regard to these abilities, it is not difficult to describe enhanced abilities simply in terms of transcendence of such limits. There are many such texts, and various manners in which cognitive abilities are elaborated. Obviously, Skinnerians, followers of Piaget, "classical" computationalists, connectionists, dynamicists have their varied ways of categorizing cognitive abilities, and the existence of a theory-neutral manner of describing such abilities is problematic (it is taken as given that there is no theory-neutral manner of explaining such abilities). Still, one may assert that there are aspects of cognitive performance that are unmistakable, and that these aspects are describable in the most ordinary terms. Reading with comprehension is one such aspect. Surely, the ability to read, say, Faulkner's *Light in August* in two hours with detailed recall of characters and events is beyond the ability of most (and the task may be made beyond anyone's current ability by significantly shortening the time). A little thought enables us to conceive similar ultrahuman abilities in memory, pattern recognition, computation, analytical reasoning, hypothesis formation, and other areas. If these categories are themselves theory-charged (folk-psychologically?) then, as has been urged in various contexts, it should be acknowledged that no speech is theory-neutral, and categorical talk of cognitive human improvement (or even categorical talk of "cognitive," of "human," and of "improvement") must be recognized as theory-dependent. While this broad concern cannot be addressed adequately here, the all-pervasive nature of such theoretical dependency (on the assumption that the thesis of theoretical dependency is correct) would appear to relegate the concern to a background status, unless special reasons bring it forward (for example, the sort of critique claiming that a theoretically united conceptual system—say alchemy—suffers overall referential failure of its theoretical terms, including those terms it takes to be commonplace).

We come to the question of feasibility. Granting that it is conceptually coherent to speak of radically increased human cognitive powers, granting that such powers would typically be of advantage to their possessors and, if generally possessed, to humanity overall, why suppose the possession of such powers is possible? If a human were cast into the open sea, it might be of advantage to assume temporarily the physiological form of a killer whale, but this ability is not feasible in the foreseeable future. Can it be supposed feasible that a human, with a human's physiology, could read a complex novel in five minutes with detailed comprehension? Simply turning the pages one by one might require more time, to say nothing of muscle limitations on high-speed scanning of a page. In this speculative discussion, we must take account of the sort of systems required to support enhanced abilities. The ability to scan a standard printed page of an unfamiliar text in one millisecond, for example, almost certainly would require enhancements to current human visual apparatus.

Surgical and pharmacological means are standard paths to enhancements in human physical condition. Regarding the task of radical enhancement of human cognitive abilities, however, such means appear limited. Genetic manipulation, microsurgery, and bionic implantation are routes that appear to offer more profound alterations than the standard paths. If we loosen our imagination, further speculative flights result—as the following passage from Paul Churchland indicates:

> . . . let us suppose that we can fashion a workable transducer for implantation at some site in the brain that research reveals to be suitable, a transducer to convert a symphony of neural activity into (say) microwaves radiated from an aerial in the forehead, and to perform the reverse function of converting received microwaves back into neural activation.
>
> . . . Once the channel is opened between two or more people, they can learn (*learn*) to exchange information and coordinate their behavior with the same intimacy and virtuosity displayed by your own cerebral hemispheres . . . language of any kind might well disappear completely, a victim of the "Why crawl when you can fly?" principle.[3]

One key to this remarkable piece of speculative bio-engineering is tricking "the normal processes of dendritic arborization into growing their own myriad connections with the active microsurface of the transducer."[4] If this last feat cannot now be done, Churchland's speculation at least isolates a problem area for further research, provided his suggestion is seriously adopted as a project. That Churchland is not making a serious proposal here should not in principle detract from his suggestion (which is made in the context of liberating us from a fixation on propositionally based folk-psychological cognitive theory).

The thrust of Churchland's speculation, in its context, is to transfer the attention of cognitive theorists to the sublinguistic realm of neural activity. But it may be enjoined that linguistic activity has been an important, if not unique, agent of human improvement. For example, new notations have, on occasion, greatly enhanced thought. Perhaps, contrary to what has been suggested here, and what Churchland is suggesting, we should confine our desire for cognitive enhancement to such as is brought about by those intellectual breakthroughs enabling more successful thought in a given domain. Even as this consideration is stated, however, we ought to be aware of the likely connection between internal enhancement of cognitive abilities and the probability of intellectual breakthroughs.

Nor will it suffice to state that we cannot in the foreseeable future realize anything like what Churchland is suggesting. For one, the foreseeable future is cloudy regarding such matters. It is enough that visionaries articulate sufficiently detailed possibilities that could form the bases of research efforts. These efforts may fail, difficulties may prove insurmountable, but such has not yet been shown.

In part, the perceived difficulty regarding the feasibility of various schemes of cognitive enhancement is due to the demand that a fully developed scheme be proposed. In this respect, the situation is analogous to that in the related field of artificial intelligence research. If an electromechanical device achieves a broad spectrum replication of human abilities, it may be the result of incremental advances in the research field, rather than a single remarkable development. Similarly, progress in neurophysiology, genetic engineering, solid-state physics, as well as other fields having some bearing on internally derived human cognitive enhancement, may lead to incremental steps towards cognitive enhancement, even in the absence of a fully developed scheme for realizing such enhancement. And even in the absence of such a fully developed scheme, these incremental realizations might eventuate in full-fledged cognitive enhancement.

We do not currently have proof that such cognitive enhancement, whether incrementally gotten or secured all at once, is possible. Returning momentarily to comparison with artificial intelligence research, Turing's famous prediction regarding the "intelligence" of computational machinery by the year 2000 has failed, and there are significant doubts as to whether it will be realized in the current century, if at all. Yet from Turing's mid-twentieth-century standpoint, it seemed likely. Such optimism predates the "frame" problem, the "grain" problem, and the general "embodied being-in-the-world" critique of Turing's project. In the light of this failure, and other shortfalls of artificial intelligence research, any proposal that serious thought be given regarding the future relations between humans and intelligent machinery seems misplaced. Nevertheless capable machinery, if not broad-spectrum intelligent, has created profound changes in the workplace, boosting productivity as it simultaneously unemployed assembly line workers. These and other such advances in machine abilities merit our attention. In some areas, such as space exploration (human versus robotic), discussion is already apparent. Such discussion can be expected to intensify as machines become more capable, failed predictions notwithstanding.

Intuitively, enhancing human cognitive capabilities in the manners indicated above is a more revolutionary project than developing machines of broad-spectrum intelligence. The former accomplishment would profoundly affect our inherent nature (minimally assuming a generalized, intuitive notion of human nature). Although we do not know how to effect such enhancements, we should examine carefully what we do know. First, we have relatively clear criteria of enhanced performance, due to our large library of detailed concepts of cognitive performance, including measurements of such performance. Second, we have hypotheses as to the sorts of physiological modifications that would enhance performance. These hypotheses include relatively well-confirmed beliefs as to the sorts of physiological effects which would underlie such advanced performance (e.g., accelerated synapse activity), vague notions regarding the realization of these physiological effects (e.g., microsurgery), and many stages of clarity and confirmation between these extremes.

Gathering the various elements of the discussion, we are positioned to address the most naturally arising criticism; namely, that it is frivolous to be discussing what are currently outlandish possibilities while confronted with the chronic problems facing humanity. One can turn to science fiction, so the criticism might go, and find a number of future scenarios that, on realization, would profoundly alter human life. And one might then make a loose feasibility case for a chosen scenario (e.g., given antimatter rocket engines with orders-of-magnitude greater thrust, we proceed to colonize the solar system. And such engines are in the not-distant future given . . .), all the while diverting attention from the fact that a large number of children in the world suffer cognitive impairment due to protein deficiency.

This last criticism carries an alternative. To what extent can we be certain that such changes, if realized, will have any real effect (not a negative effect, but any effect) on targeted problems? It may have seemed revolutionary to eighteenth-century thinkers that a person could enter their living room and experience music on demand (as Benjamin Franklin thought),[5] but having realized such we are perhaps prepared to report no general uplifting as the result. *Plus ça change* . . . expresses what many consider the effect of various ostensibly revolutionary changes. Putting the two alternatives together, speculated revolutionary alterations are often either frivolous to contemplate, or ineffective if realized.

To address these criticisms, we must acknowledge the extent to which they are correct. Urgent situations are not properly addressed by speculative flights. Technological advances are often limited to the extent that the advance is constituted merely as a new ability, without further noticeable effect (e.g., personal stereo). A morally rational plan of resource allocation would likely delegate the majority of our efforts to missions of present purpose. A rational plan of technological innovation would likely weigh the effects of a technology on a population vectored to use the technology in the best manner. It might prove impossible to develop noncontroversial rational plans of resource allocation and technological innovation, however. "Free market" capitalism has it that the most rational plan in these areas is the open play of unfettered competition. Other socioeconomic schemes offer alternatives. While we cannot enter this large discussion here, we may note that there are minimally-arbitrary touchstones. Whatever scheme has us starving in front of computer screens because we have made computers available to everyone, but have wholly neglected our agricultural infrastructure, is by most lights irrational. Using such intuitive indications of rationality, it is apparent that much of our resources ought to be directed towards current concerns.

But among the industrialized societies there has been a commitment of resources to research, and this commitment has extended to long-range research. Discoveries concerning mind/brain are potential leads to psycho- and bioengineering regarding these entities. It is in our interest to be alert to these developments, and to speculate as to their consequences. The feasibility of cognitive enhancement may emerge soon as a result of current or near future research. We would do well to consider the implications of this possibility. Further, we may wish actively to seek such enhancement through support of various lines of research.

Should radical cognitive enhancement become feasible, there will be bioethical questions regarding needed human subjects to submit to procedures thought to realize such enhancement. We may expect ethical concerns voiced against such modification of our given nature. Indeed, we have canvassed some such concerns, albeit briefly, in discussions above.

A primary concern noted above is the lack of assurance that enhancing our cognitive abilities will enhance our moral behavior. Some will have noted the absence of the suggestion that the speculated future technologies which improve our cognitive abilities be similarly employed to improve our moral capacity. I do not avoid this suggestion because I find it fatally lacking in some manner in which the suggestion of improvement of our cognitive abilities is not lacking. I avoid the suggestion because many find the specification of moral capacities more problematic than the specification of cognitive capacities. It seems to me that we can specify, to some extent, a better world in which people generally support the flourishing of other people, and not at the expense of still other people. I believe we can specify to some useful degree the distribution of personal characteristics necessary for such a world. Having specified this, we would be positioned to speculate regarding the sort of bioengineering that would yield such characteristics. But the difficulties here are undeniable. Challenges to the clarity, consistency, and nonarbitrariness of the notion of human flourishing are readily at hand. Characterizations of morality are philosophically problematic. And there have been suggestions for engineering general improvement in the social characteristics of humans which, as to means and/or ends, horrify many. Suggestions of improvement of cognitive capacities appear less troublesome with regard to such difficulties.

People can be made less harmful to others by lobotomizing them or, less radically, by maintaining them on a regimen of powerful tranquilizers. Neither of these means create citizens with productive powers. As with cognitive abilities, we currently place most reliance on socially mediated processes such as education to develop moral qualities of citizenship. And as with cognitive powers, it may arise that we decide more radical steps are needed for human survival and flourishing. Considerations on changing what we typically refer to as an individual's personality justifiably provoke philosophical reflection. Current means of doing such are limited, often with questionable side effects to the subject (e.g., electroshock treatment), and as such are typically performed as extreme measures. Were the means available, perhaps many of us would willingly submit to measures making us less envious, commodity-obsessed, egocentric, and more caring, patient, generous, and so forth. That is, we would submit given the usual stipulations regarding the vagueness of these categories, the control of unwelcome side effects, and perhaps for good measure the reversibility of the change, should unwelcome consequences become apparent.

The sorts of technologically facilitated improvement in our moral natures briefly mentioned above do not involve the introduction of as yet uninstantiated abilities. There are moral exemplars among us, and perhaps we have no clear and detailed idea of improvement on them. Rather, we wish humanity in general had similar moral characteristics to these exemplars (that there is transcultural agreement on these exemplars is unlikely, however). In this respect, the suggestion that we consider providing ourselves with cognitive abilities heretofore uninstantiated more immediately warrants the charge of tampering with human nature. Perhaps such tampering is unprecedented, but we should be aware of the extent to which we have modified the conditions of our existence in just the recent passage of time. Mathew Arnold, that eminent later Victorian, felt the change coming in human life by virtue of the industrial implementation of scientific and technological advancements, and he spoke for many when he placed himself "between two worlds."[6] Internal modifications are likely more worrisome than changes in external conditions (recall the point that profound physiological modifications would be necessary to realize certain cognitive improvements), although perhaps this worry is properly situated by noting that changes in the environment cause psychological, and hence internal (for those who believe in the internal nature of the psychological) changes.

Our future is uncertain. In our occasional thoughts regarding our collective future, preparation is not beside the point. A small part of the small part of thought we devote to preparation for our collective future is perhaps not ill spent in speculation.

NOTES

1. There is an extended discussion of the following topics in "Parallel Consciousness."

2. Leon Trotsky, *Literature and Revolution*, trans. Rose Strunsky. (New York: Russell and Russell, 1957), 256.

3. Paul Churchland, *A Neurocomputational Perspective* (Cambridge, Mass.: MIT Press, 1989), 20-1.

4. Churchland, *Neurocomputational*, 21.

5. I have tried vainly to find the source of this thought, attributed to Franklin in an audiophile publication I read many years past. Perhaps my memory is incorrect, perhaps the attribution was a mistake or fabrication. It is in any case the sort of belief an eighteenth-century thinker might entertain.

6. Mathew Arnold, "Stanzas from the Grand Chartreuse," in *Poetry and Criticism of Mathew Arnold*, ed. A. Dwight Culler (Boston: Houghton Mifflin, 1961), 185–90.

Chapter Four

Facing Immortality

We are now to speculate on some non-human possibilities. The first such speculation is that in the sometime future entities which come to replace us live non-terminating lives. The phrase "entities which come to replace us" assumes criteria of continuity over time such that relations such as ancestor-of and descendant-of can be applied to pairs of objects with little or no ambiguity. Regarding our present selves, we are the entities which have come to replace humans of the last several millennia. It is assumed that anything that replaces us, but leads a non-terminating life, is non-human. That is, it is assumed that mortality is a necessary condition of being human.

I do not insist on this assumption. One might find it prima facie plausible to imagine something much like ourselves, but living an endless life. I believe any such plausibility will vanish on closer scrutiny, but in part this depends on what is allowed by "much like ourselves." In any case, there is no clear line separating the human from the non-human. We have intuitions such that ordinary tables and chairs, as well as dogs and cats, are not human. They are surely not biologically human. It seems a stretch that any entity which could be said to have an endless life would fit the category of the biologically human, given the biological adjustments necessary for endless life. Similarly, something that lives an endless life, or at least something that believes it lives an endless life, would be psychologically at some remove from human concerns with mortality. If one insists on the possibility of such an entity being termed "human," nothing of importance in our considerations will be surrendered on granting such usage (although for various matters, such as legal contexts, much of importance can be imagined).

There is much current work on ageing. Biological organisms have obvious overall biological clocks with physiological manifestations. Roughly put, an organism grows when more cells are produced than are destroyed, an organism is at maturity when the rate of produced cells to destroyed cells is equal, and an organism is ageing when the rate of cells destroyed exceeds the rate of cells produced. If some means can be found to keep the number of cells destroyed equal to the number of cells produced for an indefinite period, then the organism will exist with arrested growth/ageing for the indefinite period of this equilibrium point. That is, it will so exist if nothing in its environment prohibits this balance.

This "balance-model" of ageing is of course an oversimplification. In complex organisms, various sorts of cells do not replace at all. And this model, at its best elaboration, is merely descriptive. Causes of such facets of ageing as selective degeneration of various tissue, replication limits on DNA strands, and higher probability of organ malfunctioning are under active investigation, at times with proposed theories in competition.[1] Still, the model serves the purpose of focusing attention on the fact that ageing in an organism is a biological phenomenon, or set of biological phenomena.

There is another dimension to clinical work on ageing. While biological research offers promise of extended life expectancy, life expectancy might also be increased by artificial means. There are various electro-mechanical substitutes for vital human organs, for example. To date, these substitutes are used virtually exclusively as extreme measures. Many of them are cumbersome, and it is only their "last resort" nature that determines their use. To speculate for a moment, heart disease is a leading cause of death in the general adult population of the United States (and those parts of the world having similar lifestyles to the United States). Imagine that all newborns receive an artificial device which duplicates all the functions of the human heart (including responses to endocrine and nerve stimulation). Imagine that this device has an indeterminately long expectancy, coupled with a virtually nonexistent failure rate, triple redundancy, and in case of the infinitesimally small probability of triple failure, ease of replacement. The onboard power supply of this device, so let us imagine, has a ten thousand year lifetime. Imagine further that the "tissue" of this device is in some manner impregnated with this power source (chemical, or perhaps nuclear), so there is no concern regarding connections of power source to device. In short, the supposition is that artificial heart replacement removes heart disease from the mortality list.

Minus other imaginative considerations, this supposition does not remove heart failure as immediate cause of death from the mortality list. Such imagined artificial hearts can fail for external reasons, such as a bullet which destroys the system. If a replacement organ is not quickly available at the moment of destruction then, other possibilities aside, the person dies. Still, it seems indisputable that such a state of affairs as speculated would by itself revolutionize human life. One's heart, or blood pump, would be removed from mortality considerations in much the manner that one no longer fears death by smallpox. Given the centrality of heart disease in current mortality considerations, lifestyle considerations, and so forth, the ramifications of such a development would be extensive.

Speculation regarding artificial hearts is speculation on the near term since, however rudimentary, artificial hearts are currently available devices. The speculation can be advanced by imagining onboard artificial kidneys, lungs, and replacements for other major organs. Some care must be exercised in this. Various bodily fluids, such as lymph and blood, are not as readily subject to substitution by artificial replacements. The complex proteins, lymphocytes, leukocytes, and so forth perform various functions not readily duplicated by foreseeable artificial replacements. Further, talk of organ replacement does not currently involve brain replacement, as we have little idea of the appropriate form of an artificial brain. Even were artificial intelligence research to provide a convincing device with broad-spectrum cognitive ability, such a device would not be a suitable replacement for the human biological brain unless it could also perform the various noncognitive biological control functions performed by the biological brain.

The last point, regarding the state of artificial intelligence research, should dampen any quick and ill-considered move to pure electromechanical robotic replacement of human beings. A future in which biological humans consciously and willingly evolve into nonbiological entities may appear a natural goal, given the sort of part-by-part replacement of major organs considered above. Yet without a nonbiological something doing much the same work as the brain currently does, visions of robotic futures remain cloudy. There is much discussion of the various problems confronting that segment of artificial intelligence research that seeks to simulate, duplicate, or possibly surpass human cognitive abilities.[2] There is also general agreement that such research has not arrived at its ultimate goals. There is substantial disagreement regarding the possibility of success of the various research efforts in reaching or approximating these goals. One of the major issues is the level of conceptualization and implementation involved in developing a nonbiological device with broad spectrum human cognitive abilities. As is well known, so-called "classical computationalists" and "connectionists" have been in apparent disagreement ("apparent" because it may transpire, as some have argued,[3] that the seeming radical differences of these approaches do not amount to much real difference) as to choice of computational model for implementing broad spectrum cognitive abilities. It should be kept in mind also that cognitive abilities are not the only aspect of human mentation.

One path towards immortality is replacement of ourselves as biological entities with nonbiological entities. Such entities are currently generally understood to be robots, by which is intended electromechanical devices, typically bipedal and in general of human outline and uprightness, constructed of metals, plastics, ceramics, and such. Differences between earlier, twentieth-century conceptions of these entities and more contemporary conceptions include some form of on-board computational "intelligence," smoother limb movement, tactile sensitivity implemented by strain-gauge technology, and broad spectrum electromagnetic sensors.[4]

To whatever extent current and foreseeable robotic research points towards self-sufficient robot beings, it is a safe assumption that the majority of humanity is not interested in evolving into such beings. What it is like to be a human cannot in the foreseeable future be approximated by what it is like to be a robot. The intuitive revulsion many humans would feel towards the contemplated prospect of evolving into robots has much to do with the belief that such entities have no mental (i.e., conscious) life. Being a robot does not differ significantly from being a rock—at least with respect to having a mental life. Further, being a robot does not guarantee immortality, given various forms of internal failure, as well as possibilities of external destructive forces.

The thoughts presented to this point suggest the following two slippery slopes:

1. Many people desire enhancement of various body parts, and replacement of defective body parts. At times the enhancement/replacement is nonbiological. The endpoint of such nonbiological replacement is a being devoid of biological parts—a robot. Yet few desire robothood for themselves.
2. Many people desire longevity. Given the choice of living at least one more year, or dying within one year, many people will choose the former. The endpoint of such desire is immortality. Yet many people will pause at the thought of living an indefinitely long life, even if this could be accomplished in reasonably good physical and mental condition.

As with many instances of slippery slope reasoning, there are replies immediately on hand. The slope ending in pure robothood is slippery only if all body parts are subject to enhancement or replacement, and only if such enhancement or replacement is in all instances nonbiological. To the first of these counters it may be replied that, given sufficient time, all human body parts are subject to failure. Hence, if there is a goal of continued existence as an integrated entity of some sort then eventual, but not necessarily simultaneous, replacement of all parts is in order. Still, the replacement need not be nonbiological. Given foreseeable biotechnologies, at least some of the replacement need not be invasive, but rather would be done within the organism, as the organism is in some manner stimulated to grow the needed tissue and dissolve the tissue to be replaced.

It follows that people need not be taken as accepting propositions implying a desire for eventual robothood. Irrespective of biological or nonbiological enhancement/replacement, however, an entity of roughly human size is subject to physical destruction by external forces. Taking the destructive potential of external forces to an extreme limit, the physical universe itself is subject to termination according to some astrophysical scenarios. Given this last consideration, it is preferable to speak of indefinite life rather than physical immortality.

It should be apparent that a major concern of those seeking replacement of defective body parts is the continuation, and perhaps enhancement, of their mental life. While there are contemplated schemes of uploading and subsequently downloading pertinent psychological characteristics, time spent in "disembodied" mental life is likely undesirable in that much of one's mental life is strongly connected to one's body. This is not to deny the value of apparent embodiment, should it be realized. Believing oneself to be embodied, if vivid enough a belief (with the usual perceived environment, including other apparently embodied persons, around one), is possibly as acceptable as real embodiment—provided the illusion lasts. There are many familiar discussions regarding such possibilities, and rather than enter these discussions let the phrase "provided the illusion lasts" be given due emphasis.

There are thus the following two distinct lines towards indefinitely prolonged human life. One line is the replacement of body parts, as needed or desired, by biological replicas or substitutes, or by nonbiological substitutes. The other line is the "virtual" or apparent embodied life offered by some sort of (computationally?) induced illusion of being in the world. Of course, the illusion line offers the possibility of biological or nonbiological replacement of bodily parts as part of the illusion. Similarly, if a biological human, or some significant biological part of such (e.g., brain) is subject to the induced illusion, the possibility exists that biological or nonbiological replacement be performed on this physical entity, unbeknownst to the consciousness experiencing the illusion.

There is yet a further line to be considered. Allowing that either biological and/or nonbiological replacement of bodily parts will not protect an individual against various external forces, or internal failures, a more radical sort of lifeline is suggested. Perhaps an analogy with a familiar computer operation is useful here. Word processing programs typically offer the user backup procedures whereby the document under construction can be saved in progress, in the event of power failure or some such threat of obliteration of the document. What if an individual's mind could be in some analogous manner subject to periodic backup or saving?

Speculations on mind storage, typically accomplished by means of computational machinery, are nothing new in the realm of science fiction. Nor has such storage been ignored by the philosophical community, there being various fanciful thought experiments of such, typically in the context of questions regarding personal identity. The idea of mind simpliciter is among the more philosophically problematic notions. Not having an agreed understanding of the term "mind," it is problematic as to what is being saved or backed up. It is often assumed that various physical configurations, commonly termed "bodies," can receive the same mind, or a continually updated version of one mind, without loss of unique personhood. This assumption is questionable, given bodily differences among these similarly minded individuals. It may come to pass, however, that a sufficiently similar body to the presumed corrupted body can be manufactured to receive the mind, in which case the problem of installation of a mind in an unfamiliar body would be avoided.

Of course, such speculation recalls the familiar problems of personal identity, such as multiple instances of the same person.[5] If a mind can be loaded in a body, either dissimilar from the one originally housing it, or a body made to be similar to the original, and if in either case the original person has a continuation of career, then it seems possible—perhaps even likely—that the same mind will occupy more than one body, thereby creating a plurality of identical selves. This and related possibilities are well canvassed in the literature. I offer only the following observations. However identically these multiple same persons initiate (imagine they start their re-embodied, or newly embodied, career simultaneously in roughly the same place—say a laboratory room), they will likely fan out and have dissimilar experiences as time goes on, thereby dis-identifying them. But they share previous memories, assuming memories are still components of minds, and this creates potential moral, social, and legal conflicts. They may each lay claim to persons and property associated with such memories. Should such a possibility of mind storage and re-embodiment materialize, there will be much to consider.

There are yet other possibilities, at least one of which need not be considered. The possibility to which I refer is the cryogenic storage of a human being, who is then revived at some later time. While such technology by itself, if successful, affords an extended—perhaps indefinitely extended—duration of a person's existence, the extension afforded while cryogenically unconscious is intuitively not a time of living. One hope of those who are willing to undergo such suspended animation is that on revival techniques will have been discovered for prolonging conscious life, perhaps indefinitely.

Further consideration should be given the "slippery slope" arguments presented above. While the slide towards robothood is apparently blocked by considerations of biological, rather than nonbiological, replacement of defective or inferior body parts, there seems little to block the slide towards wanting eternal life, given that humans typically have a day by day, if not moment by moment, desire to live. Diminution or extinguishing of this desire is generally the result of severe physical or psychological stress, such as a painful and incurable physical affliction. The question remains whether an unending, healthy life would at some point not be desired because of boredom, world-weariness, or some other psychological manifestation. This question is difficult to answer, given the lack of anyone's experience of the conditions under which the question is realized.

Since we have been contemplating the replacement of body parts in the interest of extended life, whether with biological or nonbiological surrogates, we should also contemplate the replacement of psychological "parts" in similar interest. Although speculation in this area is less grounded in genuine developments than in the case of physical body part replacement, there are examples of such. Mood-altering drugs, such as the SSRIs (selective serotonin reuptake inhibitor) and the benzodiazepines, give some evidence of the perceived benefits of replacing aspects of one's psychological disposition. In some respects, the questions associated with psychological alterations are more immediate than those surrounding body part replacement. Granted the difficulty in separating mind and body in the context of personal awareness, there is an intuitive sense that our psychological characteristics are primary in constituting human personal identity. The question arises as to the appropriateness of such psychological replacement for purposes of assuaging opposition to indefinitely extended life. This question is an instantiation of the more general question as to the desirability of a radically different psychological makeup than current humans typically manifest.

The latter question regarding psychological alteration is sufficiently large as to demand whole chapters. Prior to them, however, we will attempt a provisional response to the more specific question regarding the justifiability of changing human psychology, if necessary, so that the changed person desires indefinitely extended life.

There are few guidelines in responding to this. If all humans at all times desired such longevity (i.e., if at least the desire was present in the background of everyone's mind, if not the forefront of everyone's consciousness), such universal desire would constitute a prima facie justification of itself, barring conflicting desires. While it is undeniably true that most people, most of the time, have the desire (foreground or background) to maintain their living status, it is also likely true that most people desire to live what they consider normal or typical lives. Virtually everyone believes that normal or typical life includes a sometime termination of life, at least in the form in which it is currently being lived.

Tales of endless life coupled with unending ageing, such as the classical tale of Tithonus, have limited relevance to the present discussion, in which it is assumed perpetual withering away is not the accompaniment of indefinite life. Yet it is difficult to consider an indefinitely long life, even if presumed healthy and changelessly youthful in a physical sense. This difficulty is not solely for the reason of lack of anyone's experience. The detailed circumstances of such living are required. Is the sociopolitical climate oppressive for a long period? Is indefinitely extended life conferred on some and not others? Do the people leading extended lives watch their children, and others close to them, not leading such lives? These and other such questions need answering in conjunction with speculation on extended life.

Whatever the detailed circumstances, it can be proposed that one's psychological makeup is altered to the degree that one desires indefinitely extended life, or that one is contented with such life. Why should one, in present circumstances, desire such alteration? In approaching an answer, it should be noted that such speculated alteration is radical to the point that concern for others, including friends and family, is significantly diminished, as will be shown below and in "Alone and Without Love." In short, the life examined now is arguably not a human life, kept to intuitions regarding "human life."

Reasons can be supplied for desiring extended life under these seemingly alienated circumstances. Endless life offers one opportunity for endless experiences along the various spectrums of experiences, sensual, aesthetic, emotional, etc. Aristotle's dictum that all people desire to know is met with the observation that knowing requires time. Extended life ensures extended time. Spinoza's thought that eternal life is granted those who comprehend the timeless rationality of the universe is likewise augmented by the promise of securing sufficient time to comprehend such rationality. Spinoza, like others before and after him (consider Laplace's oversimplified deductive vision), may have thought fundamental knowledge of the universe within the thoughtful person's grasp, a matter requiring comparatively little time.

These reasons may not be compelling. However, there is the additional matter of the raw conative desire to continue life. One may think that one has no wish for the current abnormality of extended life, and that one wishes for the usual, normal, happy life of limited duration. Yet many deplore ageing, weakening of physical and mental powers, sickness, and approaching death. Human wants here are not unquestionably consistent. Wishes for normal lifespans, as currently envisaged, are perhaps clear-headed acceptance of perceived inevitability. Wants for what one believes one cannot possess, such as eternal youth, are considered best abandoned for reasons of the consequent unhappiness of nonfulfillment. All of which is sensible unless the seemingly impossible alternative is shown possible.

These considerations do not constitute indisputable grounds for the desirability of altering human psychology to the extent that humans enjoy the prospect of indefinitely extended life. Current reservations arising from common conceptions of normal life are not facilely banished. Yet neither are visionary suggestions summarily dismissed from consideration, particularly when they arise in the context of present and foreseeable scientific research.

Those desiring indefinitely extended life ought to consider general implications of such life, under any realization of it. One such result of indefinitely extended life being made available to the majority of humanity is the virtual absence of future generations. If few of the living "surrender" their supply of resources (including living space) by dying, then it is an uncomplicated exercise in mathematical reasoning to show that few more humans can be added to the earth's surface, given the earth's limited carrying capacity. While scientific and technological breakthroughs promise increased carrying capacity, there are limits to such increase.

Perhaps science and technology are being undervalued here. It may be possible to colonize the solar system, create atmospheres on various extraterrestrial bodies, even create smaller "suns," say, orbiting fusion generators issuing appropriate amounts and kinds of electromagnetic radiation. All in the interest of duplicating life on earth to a degree hitherto thought impossible. Such developments would expand resources allowing humans leading nonterminating lives to produce and raise offspring. The realization of such a vision would still have limitations, unless current physical limitations were overcome to the extent that humans had access to neighboring, and ultimately far-distant, planetary systems. Alternatively, perhaps humans could stay entirely on earth, have children, live indefinitely extended lives, and not face resource scarcity. This seemingly contradictory state of affairs could happen in various ways. One manner would be a radical physical downsizing of humans. Assume that the technology exists to downsize people to less than a millimeter average height. Further assume that the technology exists to shield people from environmental hostilities. These assumptions granted, the problem of limited resources is ameliorated roughly to the extent that it would be if humans could colonize, but were limited to, the solar system. Of course, one can double the fancy and consider the case in which humans both downsize and colonize the solar system.

The preceding discussion illustrates the extent to which our various desires are in conflict. We desire to procreate children, and we desire to live extended lives. In speculating on manners in which both desires can be realized, we somewhat unwittingly desire to preserve as much else of the status quo as possible. We imagine ourselves with similar physiques to those we currently possess, notwithstanding that they would likely not be the best physiques if we are downsized in the earth's gravitational field, or subject to other gravitational fields. We imagine ourselves raising children whose biological ageing will be arrested on adulthood, as we imagine our own ageing to be arrested, and we tend to overlook life in a society in which our children are in short time physically indistinguishable from us regarding biological marks of age. Such a society will likely have different life arrangements, different prohibitions, than currently exist. Given such changed conditions, the current and age-old desire to procreate children may have altered.

There is a literature regarding the question of our obligations to future generations, such literature being centered in the field of environmental ethics. The discussions concerning this topic all proceed under the assumption that humans have limited lifespans, as is now the case. Under such a currently realistic assumption, it seems morally reasonable to suggest that our present use and development of resources ought to be constrained by best-guess considerations as to how such resource use will impact our descendants. Generalizing somewhat, we are arguably morally obliged to take account of the effect of our actions on succeeding generations, given the usual provisos regarding the possibility of determining the results of such actions on succeeding generations. The relevance of awareness of impact on future generations with respect to the discussion of indefinitely extended lifespans is immediately obvious. The problem of appropriate social relations to offspring who are biologically of the same age as those of older generations has been noted. Perhaps more important is the question of moral permissibility of actions which likely eliminate the possibility of future generations. Granting the obligations to future generations under the ceteris paribus assumption, the question now is whether there are obligations on those leading extended lives to have offspring. Careful moral investigation may reveal that the opposite is the case; such beings have an obligation not to produce offspring.

Supposing that it is imprudent, if not immoral, for a humanity leading indefinitely extended lives to have offspring, the question arises as to the psychological impact on people who are not to have offspring. Granted that many people do not desire offspring, many do. Life without offspring may prove difficult to those desiring such. One can hope for a scientific/technological solution (e.g., colonization of solar system and beyond), allowing the creation of offspring for at least some time into a future of humans living indefinitely extended lives. Yet in the fullness of time, this option will be withdrawn unless human life is lived in radically different fashion. Here, science fiction possibilities come to mind. One such possibility was mentioned above, albeit briefly. If humans live extended lives by in some manner living "virtual" lives, then possibilities of going on in the same way increase. Imagine that humans create a self-sustaining, self-repairing (computational?) device into which individual minds can be placed such that the device will create perceptual illusions for these minds. Imagine further that for the most part these illusions duplicate the general features of physical reality, except that various sorts of real-world contradictions are overridden. For example, a population could have virtual extended life and yet raise offspring (having their own machine-kept minds) without exhausting resources in that the machine would make perceptual adjustments from each individual's perspective so as to present the view of sufficient resources for that individual and those with whom the individual is involved. Unfortunately such a scheme, even if possible, is likely not one to which humans would knowingly accede.

If we contemplate a radical changing of our bodies, various extent psychological generalities applicable to humans, commonly considered desirable, may no longer be deemed necessary or even desirable. As some sort of robots, for example, various of our current affections and appetites will likely disappear. We may have little or no regard for one another. We may lose our desire for fine cuisine, satisfying sexual activity, and so forth. Given that having regard for one another, forming friendships, enjoying forms of satisfaction of various appetites, and other related activities constitute a significant part of what many think gives value to life, an extended life without such enjoyments is seen as not worth living.

Yet we desire extended life—at least many of us desire to live long lives in good health, and to postpone biological ageing. As has been shown above, the common desire for extended, healthy, youthful life is naive, in that closer examination reveals incompatibilities in the various common desires of current human life when these desires are set in the context of extended life. The problem here does not differ much from the problem of the naive wish on the part of many human males to be bigger and stronger than they are. Unless limits are placed on such desire, one's fancy may run to goliath-like proportions, unaware of the severe strains such size places on the load-bearing joints of the standard human physique. Even as there is a proper range of sizes for the current human form, errant wishes aside, so there is a proper range of lifespans for the current biological, psychological, and sociohistorical human.

If one is enough of an ethical egoist, one might try the position of indefinitely extended life for oneself, and perhaps a limited number of others, while denying it for the greater part of humanity. Such a position does not fall to the previously cited resource objection, should the near-immortal few desire to raise children (who may themselves be given extended life, or perhaps denied it). At least, such a position does not fall as quickly to the resource objection.

The situation of offspring who are themselves near immortal, and who themselves have near-immortal offspring will, in the fullness of time, realize the objection. It is a question of numbers as to the nearness of the difficulty in such contemplated circumstances. More telling is the psychological alienation from the majority of humanity the "extended ones" are likely to experience. Such psychological effects are of course not logically necessary, and may be avoided in various manners. Should they be avoided by significant alteration of common human psychological constitution, however, then the questions regarding what sort of entity to which we are conferring extended life reappear.

The discussion has proceeded on the assumption that the common human wish for extended life ought to be acknowledged. This wish has been shown to be in apparent conflict with other human desires, such as the having of offspring. On the further assumption that humans gain increased control over various aspects of their psychological makeup, it can be hypothesized that the desire for offspring is modified, perhaps eliminated, in favor of the desire for longevity. But if mastery of psychological characteristics extends to the tempering of desire for children, so it might be argued, why cannot such control extend to the desire for longevity? There are desires that humans are arguably better off without. Perhaps the "fountain of youth" desire—to paint it ungainly—is such a dispensable desire.

The general form of the problem raised has a history in philosophy. From what standpoint, assuming a variety of such, do we measure the comparative worth of our desires? The uneducated (in some broad, intuitive sense of "educated"), popular-media-drenched person may have intense desires not shared by the cultured (again, in the broad, intuitive sense) person, and conversely. J. S. Mill was sufficiently thoughtful so to query the presumed universal desire for pleasure, thereby contradicting the purely quantitative hedonistic calculus of his predecessor, Bentham. The reasoning supporting the answer Mill gave is as old as Plato (*Republic* 582); the person with the experience of the various possibilities is the one to decide. Mill thought such a one would prefer a life focused on the higher pleasures of the intellect rather than the lower pleasures of the satisfaction of bodily appetites. Of course, there are other conceivable principles of adjudication which would not support the answer Mill was clearly seeking to support. Further, will any principle that adjudicates "higher" and "lower" desires unequivocally decide between the competing desires of longevity and procreation? Is one of these two desires obviously "higher" or of greater moral worth than the other? There may be a temptation to respond that the desire to procreate and nurture offspring is intuitively of a higher moral order than the self-centered desire to live an unending, biologically youthful life. However, at least in part this response assumes that the desire for longevity is not the desire for everyone's longevity. Indeed, this desire has not always been universally generalized (how likely was Ponce de Leon's quest for the fountain of youth inspired by a wish that all, and not just the Spanish nobility, partake of its waters?). It is also likely that the desire for youthful longevity is not infrequently conjoined with desires for interminable bodily satisfactions, so that what is being morally weighed is the desire for endless sexual gratifications, without obligation to future generations, against the sober requirements of child rearing in a world of limited human lifespan.

The desire for longevity can obviously be colored differently than has just been done. One might even consider a retort to the notion of "the sober requirements of child rearing" in terms of hideous intended ends for the reared children, so to balance the former picture morally favoring child rearing with one morally favoring universal longevity, a universal longevity, further, in which participants pursue intuitively noble goals. There is nothing new in the observation that intuitions based on facile sketches are facilely manipulated, and it is well to keep this observation in mind while engaged in the sort of deliberations occupying us here. The caution granted, it is difficult for the supporter of the worth of longevity over the worth of procreation to defend such a position against a claim of the following sort. Once extreme characterizations are dispensed, the desire for longevity has a general air of self-serving that is less apparent in the desire for offspring. There is a clear sense of self-gratification in having children, and some people have an intense desire for children that has little to do with thought of the perpetuation of the human race. Still, having children has a prima facie component of other-regardingness that is prima facie lacking in acting so as to live an indefinitely extended life.

At this point in the deliberation of endless human lifespan without children, versus temporally limited lifespan with children, the scales seem somewhat tipped in favor of the latter. However, this is due to the intuition that the desire for children is morally more commendable than the desire for indefinitely extended life, various other matters remaining standard. The ceteris clause here should be studied carefully, given the radical nature of the contemplated alternative of extended life without offspring. It is likely that in our wish to avoid radical change, we favor a going-on-in-the-same-way, and find support for this in the moral worth of having children over self-servingness. We are by ontological definition choosing from the standpoint of the status quo, which is to say that our choices are our choices as sociohistorically situated persons. Nevertheless, because one of the alternatives under deliberation is outside the status quo, we should be alert to moral intuitions attached to the present general circumstances, which circumstances are typically perceived as welcoming children. There are situations in which considerations of common welfare outweigh the seeming moral imperative to value having children over not having children. Communities with strictly limited means, such as Eskimo communities of earlier times, may have needed to limit the having of children. The government of present-day China places restrictions on the having of children, such restrictions being justified by concerns for the common good. If an understanding of the common good includes indefinitely extended life for all, then the continued general having of children will apparently be seen as contrary to the common good.

If there is a decisive consideration enabling us to adjudicate between our desire for extended life and our desire for children, it is elusive. Rather than continue the search for reasons settling this dilemma, let us investigate reasons for the radical alternative—extended life without children. The proposal is that, provisionally, the following attitude be adopted. The suggestion that humanity abandon its age-old desire to rear children for the sake of indefinitely extended life is on first hearing sufficiently repugnant to many, so that it is incumbent on anyone presenting this idea to offer something on its behalf. If nothing convincing is forthcoming, the suggestion that humanity forego having children can be dismissed as the outrage it initially appears to be.

The manner in which the case will be made for choosing extended life over the having of children is by means of comparative social sketches. Various forms of human life will be presented in brief description. All such descriptions will be of present or future possibilities (feudalism, for example, will not be considered). It will then be argued, on the basis of these descriptions, that the best of them is that which includes indefinitely extended life. Which description offers the best life will be determined by our present wants and values. It should be understood that what is being offered is a best life for most people, rather than for a minority. The presumption is that there are common wants and values determinative of a choice among these descriptions.

The most familiar of these sketches is here termed the "status quo sketch." What is imagined is human life as occurs now, with large population areas having material well-being and large sectors of material impoverishment. There are periodic regional wars, but the threat of nuclear escalation typically leads to efforts among the nuclear powers to limit such conflicts. People in industrialized societies are generally not threatened by famine or pestilence, and such disappointment or sorrow as exists in their lives may be thought to be self-manageable, even if particular selves fail in some manner to manage it. In this sketch life goes on as it currently does in the real world.

The second sketch is termed the "progress model." This model is grounded in the perspective that progress is improvement, and a rough measure of progress is material prosperity—such prosperity understood intuitively. A further part of this model is the tendency towards egalitarian distribution of prosperity. In this model, the populations of underdeveloped sectors, including those pockets of underdevelopment in developed societies (again allowing common intuitions regarding development-underdevelopment) are raised to the level of material well-being existing among the developed. Such material well-being includes concomitant levels of education, health care, and recreation.

The third sketch is termed the "devolution model." Herein the social task is to simplify, in the intuitive sense of the term in the context of this discussion. Put thermodynamically, total caloric throughput is lowered. People travel less, live in smaller houses, are subject to more preventative medicine and less "extraordinary" medical procedures. Common diets shift from an emphasis on the top of the food pyramid to more vegetarian fare. Somehow ambition is diverted from the common desire to sit at or near the top of the corporate world (which world itself is in some manner de-emphasized), towards the desire to promote community well-being, such communities being scaled down from "megalopolises."

Last in this series of sketches is the depopulation model. While easily teamed, or conflated, with the devolution model, this model is logically independent of the devolution model in that the prevailing assumption of this model is that worldwide population is lowered and controlled so as to remain lowered. While there are drastic means of accomplishing this end, let it be granted that only the most humane methods (birth control, educating populations to have smaller families, etc.) are employed. Let it further be granted that population control is done in an egalitarian manner, so that the wealthy—if there is a wealth disparity—do not feel privileged to have larger families than others.

There are of course many other conceivable social sketches. The four given here are representative of those of morally good will. It is assumed that those who offer any of the four above are well intentioned, in the sense that they have the general welfare in mind, rather than desiring personal advancement or the welfare of a selected group at the expense of others. Admittedly, matters are not so simple as expressed here. Those arguing for self-interest, resuscitating social Darwinism, dividing the world into believers and infidels, and so forth, typically have some sort of moral theory at the foundation of their thinking. Still, it will be assumed here, without further argument, that schemes favoring universal general welfare at the outset are morally preferable to schemes reasoning that selective benefit occasions long-term egalitarian benefit.

There are familiar debates as to which of these four sketched societies is preferable. Environmentally oriented people tend to favor devolution and/or depopulation schemes. Those thinking environmental concerns need be balanced with requirements for human flourishing often favor schemes such as that of the status quo, or the progress model, or some social scheme lying between these two. There are well-known advocates and manifestos of these and related schemes. The reader likely has his or her favorite scheme either consciously in mind, or operating as a background to specific positions taken. While the sincerity of the advocates of each scheme is, at least in some cases of advocacy, unquestionable, those supporting other schemes have ready arguments that move to implement or maintain another scheme promote an unacceptable amount of misery. The choice of a social scheme, including maintaining the status quo, being unavoidable, how should humanity proceed?

The laissez-faire inclination of some manifests itself in the response that different societies should proceed according to their own ethos. By what right, it is questioned, do we choose the appropriate mode of society for the outback aborigine? By what right could the aborigine choose for us? While this is an extreme instance of the general question as to who chooses the sociopolitical structure, several current sociopolitical conflicts around the globe show the vexing and pressing nature of the overall question. Perhaps the outback aborigine, or the Amazonian Indian, can escape the pressures of social change—at least for a while longer. However, every time someone gets into an automobile ("I have to work, and I need a car to get to work"), turns on an electric light, or performs any of a myriad of daily tasks, that someone is adding their support to constraints on the range of social choices generally available to humanity. In living our lives we are not merely showing our choices; we are supporting particular forms of scientific-technological, productive-distributive systems, backed by political systems capable of militarily safeguarding the overall system, which is largely a global system. There are occasions when these obvious points bear repetition. There will be those dismayed at the "might makes right" flavor of the points. Perhaps some of this vexation is assuaged by the observation that it is not simply the might of the overall system, but also the continuing survivability of the society implementing it, that determines its upholders right to legislate its propagation.

If the question of choice of social scheme is phrased not in terms of mutual toleration of all forms of societies, but in general terms of overall happiness of populations under one or the other of the four sketched schemes, then the "might makes right" sort of response is unhelpful. While the point regarding continued survival might be of some use in such deliberation, it will be allowed here that advocates of each of the four schemes can make a prima facie case for social reproduction under their scheme.

Let the discussion proceed in the following manner. Let it be assumed without argument that the best of the four social sketches; id est, the system most of humanity would prefer if apprized of all relevant facts and implications therefrom (this vague but intuitive standard of value is offered here without needed support), is the progress model. Given the absence of supporting argument, the reader is under scant rational compulsion to accept this choice of model, or to accept the consequences of this choice. There will be consequences, for it will be argued, on the basis of this choice, that humanity's best interest includes pursuit of indefinitely extended life.

Such an argument for extended life is subject to the objection of question-begging. If the childless consequences of indefinitely extended life are sufficiently negative, then if these consequences are based upon the choice of the progress model, perhaps the progress model should not be chosen. If the argument from the progress model to the choice of extended life can be made regardless of which social sketch is chosen, the objection of question-begging is avoidable. However, the argument cannot be made. As will be seen, the argument for extended life from the choice of the progress model would become an argument weighing against that choice, in view of its childless consequences, were any of the other three models of society adopted. Even allowing the cavalier assumption that the progress model is the best of the four competing social sketches, if one of its implications is a nonprocreative future for humanity, perhaps the progress model should only be held to be best in itself, apart from the consequence of childlessness.

Again, without attempting any justification of the value placed on the progress model, the support for its choice by humanity is simply this: it will be humanity's choice. The evidence for this statement is, first, it has been by and large humanity's choice. Most people in the modern world have chosen what is here broadly termed "development" over underdevelopment, their desires indicated by their actions. Second, there is nothing unexpected in this choice. People are generally so intellectually constituted as to realize that a certain level of material prosperity is a necessary, if by no means sufficient, condition of living well. Those of the developed sector who exhibit a certain world-weariness, or full-fledged contempt for material prosperity, tend to ignore that such prosperity has been a necessary condition of their receiving the intellectual development (i.e., education) from which they disavow the conditions allowing such development.

The progress model assumes that the majority of humanity can have a standard of living currently enjoyed by well-off members of the developed world. There are well-known arguments against this assumption, such as those predicated on resource scarcity. It will be largely assumed that such contrary arguments have been, or can be, met. As a softening suggestion to the antipathy generated by this apparently blithe assumption, consider one case presented by opponents of the assumption. The point is occasionally made that if the various underdeveloped populations of the world lived automobile-dependent lives to an extent similar to those among developed societies, pollution and raw materials consumption rates would be disastrous. Such criticism has its own assumptions, such as that transportation technology will remain roughly in stasis. If certain cleaner forms of internal combustion are developed, such as hydrogen burning engines, foreseen levels of pollution may not materialize. If different forms of mass transportation are developed, such as high-speed magnetically levitated railways, intermediate and long distance travel may be possible at reduced materials cost. Shorter distance personal transportation might be done in a manner other than with large automobiles.

This brief point and counter is of course not decisive. In its non-finality the discussion shows the need to make bold what-if assumptions if speculative matters are to be entertained. Granted there may be unbearable consequences in raising standards of living of the worldwide materially dispossessed. It may be worth exploring the implications of assuming there are not necessarily such consequences.

Material well-being has not necessitated satisfaction. Aside from psychological problems attending worries of loss of such well-being due to relevant economic events, those secure in their material well-being are occasionally dissatisfied with their lives. Such dissatisfaction arises from a multitude of causes: difficulties with family or friends, perceived deficiencies in social status, need for further material acquisition, and so on.

Those leading self-satisfied lives in materially well-off circumstances (i.e., those who do not desire significant change in the sort of lives they lead) may yet be leading lives many would consider lacking. It is a disquieting fact that a significant amount of time spent on the world wide computer network (the internet) is for the viewing of pornography. The United States may claim the largest population of materially well-off citizens, but it also lays claim to the greatest proportion of those severely overweight, and for the most part such excessive caloric intake is not comprised of haute cuisine. Thomas Carlyle railed against democracy as "free reign in the cheap and nasty." Perhaps his criticism was misplaced. Whether or not people have a voice in their governing, having economic access to the sorts of surplus generated in modern industrial societies, lacking censoring controls on items produced, is seen by some to realize Carlyle's fears. In sum, there is arguably a malaise attendant "the good life" when this life is largely defined in terms of the transient satisfaction of the appetites.

Raising standards of living of those materially impoverished is not in itself all that is the case. Accomplishing this task, and not merely giving it the lip service world leaders and spokespeople for dominant economic institutions currently give it, would likely engage the consciousness of large segments of populations in developed societies. The alienation of living in separate societies, both in regard to other nations and other societies within one's own nation, would be mitigated. There is hence reason to suppose that the achieving of significant economic unity amongst humanity would have profound psychological consequences disseminated among many.

If this optimistic note does not appear sufficiently naive, let it be added that were humanity at this point people might collectively wonder, as they rarely do, what is to follow? People are typically sufficiently occupied with personal, familial, neighborhood, or community concerns so as not to wonder where humanity in general is headed. Yet we are supposing a remarkable state of affairs in which the overwhelming majority of humanity is well fed, well housed, literate, with access to competent medical care and, in keeping with all this, aware that this has been accomplished for everyone. Granting this, it is perhaps not a large stretch to imagine worldwide concern for humanity's future.

It is impossible to predict with assurance how most people will be regarding the future of humanity at the proposed point of universal material well-being. As noted, such thoughts might be absent most people's minds. There is an additional consideration, however, that has not yet been brought to bear in this discussion, and this consideration is likely important. In imagining the universal distribution of material prosperity, as such prosperity is currently understood, optimism for humanity's prospects under this prosperity is constrained by thoughts of misuse. If people's material well-being is assured, the concern that people will lapse into self-regarding hedonism appears a probable and disturbing outcome. While such hedonism need not take the form of imperial Roman practices, as reported by Seutonius and others, it will strike some that dramatically raising living standards for the currently impoverished people of the world so that these people, along with all others, can enjoy the vapid pursuit of pleasure, is not a fitting goal of the struggle to have arrived at that point. But in imagining this ostensibly unsatisfactory turn of events we have not taken account of the discussion of the previous chapter.

The broad nature of the social sketch allows large latitude of detailing. Among the possibilities, consider that people's cognitive abilities have been raised in the manner presented in the previous chapter. To wit,

> Consider the expectation that children before adolescence be literate in science to the extent that organic chemistry, relativity and quantum theory, molecular biology, and so forth are understood by them to such degree that they can routinely solve the sort of exercise problems currently given in texts read by graduate students specializing in these various disciplines. That is, the average child manages

this for all these disciplines, as the average child today manages adequate performance in arithmetic, basic reading skills, basic knowledge of national history, geography and so forth.

As this speculative "average child" matures, it learns further scientific matters, enhances its knowledge of the humanities, social sciences, and so forth.

If we imagine the average human in this cognitive position, the probability of people choosing a life of mindless hedonism appears remote. Granted, as long as humans have appetites and derive pleasure from satisfying them, pursuit of appetitive pleasure will not be absent human life. As Plato, and many following him, have noted, such pursuit is to a greater or lesser extent at the expense of moral, social, and cognitive development. Perhaps Conrad had it right in *Heart of Darkness*. Perhaps the urge to "go native," to "revert," lurks in the subconscious, to be released in the right circumstances. Yet if Conrad's Kurtz, the cultured philanthropic European who turns to primitive appetites when removed from civilization is any one of us, as Conrad intimates, then one manner of defense against such reversion is not to be removed from civilization and placed in the company of "savages." This defense is assured if humanity is developed to the point where savagery does not exist. But, to stay with Conrad a moment, there is the savagery of the uncivilized, the "native," and there is the savagery of the civilized, the colonizer for example. Various events in recent history attest to the reality of the savagery of the civilized.

There has been a drift in the discussion from the danger of hedonism as an available choice for the materially well off to the danger of savagery erupting among the civilized. Either of these possibilities may follow on the sort of cognitive development sketched above. The hope that this will not happen is grounded in the cooperative effort needed to exploit fully the cognitive capabilities which humans have brought to be. The dissemination of information, coordination of scientific endeavors, and general enhancement of intellectual life mitigate against asocial hedonism and uncivilized savagery.

If humans, as presently considered, develop cognitive capabilities such as those outlined in the preceding chapter, then they are likely to desire the extended longevity necessary to use these capacities both to absorb what has been discovered, and to make new discoveries. Put another way, a great sorrow of humans is alienation. In addition to the forms of alienation familiar to us from Marx, there is the alienation from reality itself that is ignorance, an alienation Plato saw at the root of wrongdoing (even if Plato did not express ignorance in terms of alienation). Granted, Marx would likely consider the notion of ignorance-as-alienation-from-reality to be a bourgeois perversion of his notion of alienation attendant the social relations of production in historical and contemporary society. Yet there is in the notion of alienation the idea of an ameliorable ontological dispossession of the human from some positive value. Humans have a cognitive nature that some thinkers have pronounced definitive of the species. As cognizers, humans seek a unification with reality through knowledge of reality. As cognizers, ignorance is a displacement of ourselves from what could be ours.

The suggestion that indefinitely extended life promises ever further knowledge encounters strong philosophical objections. If one accepts the general outlook of thinkers such as Thomas Kuhn, one finds the quest for ever-increasing knowledge of reality to be fundamentally misguided. According to the view of Kuhn and related thinkers, there is no final theory (or "paradigm"), or set of theories, under which phenomena can be fully understood. Rather, what counts as items of interest, as well as the very nature of phenomena, depend on historically situated theories, which will endlessly evolve in unexpected manners. Hence, knowledge—or at least knowledge of natural phenomena—does not accumulate in the long run, but only in the comparatively short run of an accepted outlook. When that outlook changes, so does what counts as knowledge. So does what counts as reality. In a manner of speaking, human understanding starts over when theories are replaced, but with an appreciation of where it has been, and typically with increased sophistication in technique.

While this is not the place to argue the overall merit of Kuhn's, and related, views, it should be noted that to some extent these views can be accommodated here. We can attempt an argument, contrary to Kuhn, that knowledge is largely cumulative, and that as humanity temporally progresses, more is known, and the quantity of what is unknown decreases. Such an argument premises no periods of devolution, no "dark ages." It also premises a fixed quantity of items of knowledge, and likely premises a notion of knowledge as entailing accurate representation of reality. Various influential discussions in the latter half of the twentieth century, some with roots extending to the earlier part of the nineteenth century, cast doubt upon these premises. However, there is no need to reply to this critical movement in maintaining the cognitive benefits of indefinitely extended life. Let it be granted that knowledge is somehow relative to sociohistorical context, and that coherentist and/or pragmatic epistemologies are superior to more conservative, nonrelativistic, correspondence-based epistemologies. In whatever manner knowledge is characterized, as long as it is allowed that in some manner it exists (contra radical epistemological skepticism), one's cognitive experience is partially a function of one's lifetime—provided one's cognitive faculties are active for the duration of one's life.

Perhaps this accommodation to those suggesting that the philosophical waywardness of hopes for endless cognitive activity, which are based on conservative epistemological premises, is too quick. Surely, some of the attractiveness of lessening the alienation from reality that is ignorance is rooted in questionable assumptions of what is being termed here "conservative epistemology." It might be added that questionable conservative ontological assumptions are meshed with this conservative epistemology. The reality we are supposedly alienated from in our ignorance seems one favored by an ontological realist, rather than, say, a Deweyian pragmatist. Items of this ostensible reality are amenable to our knowing and exist in a definite state independent of our cognitive activity regarding them.

Again, the issues surrounding ontological realism cannot be adequately discussed here. But perhaps some bluntness is in order. There may be those for whom the denial of ontological realism and the so-called conservative epistemology associated with it is largely an excuse to abandon the labor of learning. If there is nothing really to know, and no manner of really knowing it, then scientific activity is arguably best seen as culturally-based political activity of some sort. If this last remark strikes one as callous and superficial, perhaps we would be best served speculating on the manner in which Dewey or Kuhn would regard the opportunity to engage in scientific thought for an indefinitely extended period. For all the denial of ontological realism in their philosophical outlooks, can there be much doubt as to their enthusiasm for the opportunity to remain associated with the flow of scientific thought as it (in their views, endlessly) unwinds?

And for those for whom extended opportunity to learn appears an onerous task indefinitely extended? As human matters currently stand, there is little to be said in opposition to their view. That is, if the opportunity to learn seems boring to them, then it is such to them. If one thinks one is bored, one is likely bored. Others can attempt to persuade a bored individual that they should not be bored, that involved states of mind are more attractive than bored states, but if the bored person remains bored, so be it. As things currently stand, most healthy adult men are not bored by the prospect of sex. Most children ages seven to ten are not bored by the prospect of a visit to the toy store or its updated equivalent, the electronic game store. Most, or at least many, people in active retirement in the developed sector are not bored by the prospect of travel. If the sort of cognitive improvements suggested in the last chapter were implemented to a significant extent, many people—if not most—would arguably not be bored by the prospect of endless learning. As matters currently stand, the desire for travel is to some extent a sociocultural phenomenon. The impetus to travel occurs among some, at least partly, due to expectations as to what one should want. If one is of a certain socioeconomic status, one is expected to travel. Various present desires are alterable in the foreseeable future. If as a species our cognitive powers are enhanced, it may well result that the cultural climate is altered to the extent that learning assumes more fashionable garments.

To speak of altering desires is to sound an ominous note. The topic is unavoidable. It is connected to the questions that have been raised in this chapter. If human life can be indefinitely extended, do humans facing that prospect want to continue indefinitely in their current form? Would a childless future be less undesirable in an altered form? What if indefinitely extended life demands an altered form? Can such alterations as provide indefinitely extended life be largely restricted to biological alterations?[6] Suppose that indefinitely extended life involves a profound transformation of the current biological nature of the species. At best, we are wary of such change. Yet extended life is not without attraction.

To move off the impasse, let us assume that such changes are made as are necessary to enhance the desire for indefinitely extended life. To the extent that a return to previous conditions is possible, let us sweeten the arrangement by supposing the means to restore humanity to its current biological status, should such restoration later be generally deemed desirable. Of course, what many would now view as reason for such return—on a showing of consequences—might later not so appear once alterations are effected.

Where might we go next?

NOTES

1. For a survey of theories see Marie-Françoise Schulz-Aellen, *Aging and Human Longevity* (Boston: Birkhäuser, 1997).

2. See John Haugeland, *Mind Design II* (Cambridge, Mass.: MIT Press, 1997) for examples of pro and contra discussion.

3. For example, Jerry Fodor and Zenon Pylyshyn, "Connectionism and Cognitive Architecture: A Critical Analysis," *Cognition* 28, no.1–2 (March 1988): 3–71.

4. For lively discussion of these and other possibilities see Chris Gray, *Cyborg Citizen: Politics in the Posthuman Age* (New York: Routledge, 2001).

5. The most amusing, and among the most provocative, discussion of such problems is Daniel Dennett's classic piece, "Where Am I?" in Dennett, *Brainstorms: Philosophical Essays on Mind and Psychology* (Cambridge, Mass.: MIT Press,1985), 310–23.

6. See Stanley Shostak, *Becoming Immortal: Combining Cloning and Stem-Cell Therapy*. (Albany: State University of New York Press, 2002) for an extended discussion of this possibility.

Chapter Five

Parallel Consciousness

> This fact at least raises the possibility that much (most?) of the parallel activity in the brain also has nothing to do with active processing but is required for maintaining stable memories. The evidence is not available for settling this question one way or the other, although it is worth pointing out that an increasing number of magnetic resonance imaging (MRI) studies show, during the subject's performance of a task, rather localized brain activity that migrates from one part of the brain to another in a matter of seconds. This would usually be regarded as a symptom of serial activity. —Herbert Simon

> One might assume, given that parallel processing takes place in search tasks, that our perceptual machinery simultaneously builds a representation of the location and properties of multiple objects. If this were the case, people ought to have certain abilities that they in fact lack . . . The fact that we cannot do these things challenges our naive assumptions about perceptual experience and raises interesting questions at the border of experimental psychology and the philosophy of mind. —Harold Pashler

I want to explore a general manner in which human intellectual capabilities might be improved. This betterment has little possibility of near-term implementation. Nevertheless, it will be of some philosophical profit to explore the terrain opened by the proposal.

The phrase "human intellectual capabilities" invokes a plethora of general to specific types of mental dispositions, not all of which fall under my proposal. By the phrase I intend such aptitudes as generalization, abstraction, association, and hypothesis formation. This brief list can be cross-categorized by lists such as problem solving, pattern recognition, navigation; or understanding, memory, speech, and so forth. All the mentioned items, and many more unmentioned, fall under the proposal to some extent.

These capabilities would be improved if their associated functions were done in less time, all else held constant. Such improvement might appear initially to be of small consequence. There are circumstances, however, in which a gain in speed of thought has significant value. Anyone who has struggled with the details in comprehending an established mathematical proof or scientific theory appreciates that the faster those details are grasped, the better the chances of an overall in-depth grasp of the proof or theory. It is often through repetitive reading, together with the separation of the presentation of the problem or theory into manageable parts, that such pre-established results are comprehended. Given the time-related decay of short-term memory, it is likely that repetitive attacks on the problem facilitate a shorter time frame in which the details can be held in the forefront of consciousness, thus aiding overall comprehension.

There are of course other important circumstances requiring our intellectual capacities besides the comprehension of established results. In forming original hypotheses, for example, we often desire that such hypothesis be profound, and not merely quickly attained. Still, a significant speedup in the generation and thought-testing of hypotheses increases the likelihood of profound discovery, all else being equal. Further, hypotheses formation necessitates previous assimilation of relevant information, and the time required for this prior process often bears upon the success of the hypothesizing.

Having briefly indicated a possibility of improvement in human intellection let us briefly consider the favorable consequences of such improvement. Human thought, applied to the task of heightening the material conditions of human existence, has yielded long-term solutions to various problems confronting humanity. A historical survey of our modes of production indicates that we have created increasing surpluses enabling proportionately more of humanity to engage in intellectual labor, which labor has at times been applied to the creation of conditions promoting further surplus. While it cannot be maintained that an increase in material production is sufficient for the overall improvement of life, it is arguably correct that such increase is necessary if a majority of the world's population are to lead lives generally thought better than those currently lived. Should this reason for undertaking a radical improvement of human intellectual capabilities be rejected, there remain other reasons, such as that of individual satisfaction, for favoring the improvement of such capabilities.

How might we improve our intellectual capabilities? One answer has been given: by thinking faster. Other intuitively correct answers to this question are readily at hand; "more reading," "more sleep," "improved nutrition," and so forth. These somewhat simplistic responses are easily supplemented by more interesting suggestions. Some such further replies regard socioeconomic reform, others concern conceptual revolutions. Other deep replies are conceivable.

One response to the above question, which appears less to be deep than to be wayward, is that our intellectual capabilities would be improved if the manner in which we performed our individual intellectual tasks were altered from a basically serial process to a basically parallel process.

As these terms will be used below, a "parallel process" is a process in which various subtasks of an overall task are performed concurrently, the results of these subtasks ultimately being integrated. A "serial process" is a process in which the distinct subtasks of an overall task are performed in succession, the results of the subtasks ultimately accumulating as an overall result. Among the problematic notions in these two characterizations are "task," "subtask," "integration of the results of subtasks," and "accumulation of the results of subtasks." "Tasks" are understood to be goal-oriented processes. "Subtasks" are understood intuitively as the articulate partitioning of tasks into smaller component tasks. There is a further reliance on intuitions regarding the notions of "integration" and "accumulation" of subtasks. These are important notions in the field of computer science, so that reliance on intuitions should not be taken as minimizing their importance. In particular, there is a significant and growing body of work on parallel processing. However, the unavoidably technical aspects of this work should not cloud the obvious thought that for parallel processes to culminate in an overall result, their separate contributions must be combined in some manner.

The term "consciousness," appearing in the title of this study, is philosophically problematic. The term will be used here with the expectation that questions of dualism versus monism are avoidable. Rather than take the extreme physical-monistic position that terms such as "consciousness" refer to nothing, let us adopt the relaxed view that, broadly speaking, humans are sometimes conscious in their activity, and sometimes not. This includes the possibility of shades of consciousness from, say, "dim-awareness" to "sharp-focus." Whether any such consciousness is ontologically separate from concomitant neurophysiological activity is a question not to be pursued here.

The term "consciousness" is used here by reason of convenience borne of familiarity, but not solely for this reason. Indeed, the history of science is littered with terminology and notions long discarded as being nonreferential, or misleading, or of no scientific import. Our seeming consciousness is perhaps similar to the seeming necessity of Euclidean space. Rather than abandoning the notion of consciousness, however, it will be employed in this study in as ontologically neutral a manner as possible, for the reason of its usefulness as a contrast to the lower-level descriptions of brain activity.

It is generally accepted that the brain is a massively parallel device, in that many of its operations are simultaneous. This is apparently true at both the neurophysiological level and the functional level. Brain scans reveal concomitant activity in widely distributed areas of the brain. Viewed as a regulative/control/executive device, the brain is performing various functions at what appears to be the same time. It is of course possible that such ostensibly parallel functioning is really being accomplished serially by rapid switching, much in the manner that sophisticated time-sharing computational architecture is apparently parallel, but actually serial. To the extent that the various ostensibly parallel process/control functions can be connected to the genuinely parallel neural activity distributed throughout the brain, this objection can be set aside. It should be noted, however, that computers show much concomitant distributed electrical activity, even as they carry out what at the programmatic level must be described as serial processing.

Phenomena referenced as "conscious human thought" afford an apparent contrast to the likely parallelism of much brain activity. In performing various cognitive tasks, particularly those requiring conscious deliberation, humans proceed in what seems to be step-by-step serial fashion. While some deliberative tasks inherently demand serial thought, in that some stages of the deliberation cannot be addressed until others have preceded, there are many imaginable cognitive tasks in which it is conceptually, if not psychologically, possible to divide the task into subtasks that a group of people could work on separately and simultaneously, integrating their results at some point. What is of interest here is that there are many cognitive tasks of which it seems no one person can perform such simultaneous thought.

The literal context of "parallel" and "serial" process is electrical circuitry. Held against this context, much of the use of this terminology in other contexts is imprecise. Conscious processes, for example, could be considered under some sort of path-ordering principle stipulating that such discontinuities as wayward associative drifts, or nonassociative "random" startups, negate seriality. If such cases are so considered, introspection might reveal that most conscious thought is not serial for any significant duration. In comparison, a group of physically discontinuous electrical circuits do not constitute an overall serial circuit, even if humans sequentially activate these circuits. Still, there is reason to regard at least some sorts of consciousness as serial. Those instances of conscious thought which are goal directed, for example, often have a starting point and an endpoint. During the thought process of problem solving, there are occasional discontinuities, after which the main stream of thought is resumed. Held to the literal electrical understanding of "serial," circuits are easily imagined which are casually understood to be serial, albeit discontinuities occur in their overall path. Alternatively, if the strictest understanding of "serial" is demanded, then it is unlikely that the term will be of use in characterizing mental processes. For even allowing the existence of moments of conscious thought proceeding without apparent break, the differences between the flow of electrons through a wire and the succession of somehow-connected conscious thoughts are significant.

In regarding brain activity at the neural level as parallel, the comparison with electrical circuits might be literal, rather than merely analogous. However, a discrepancy between the initial definition of "parallel process" and the electrical notion of a parallel circuit should be noted. The initial definition of "parallel process" was given in terms of concurrent subtasks. But not every division of an overall process into concurrent sub-processes is divided according to tasks and subtasks. Some parallel elements of electrical circuits function to accomplish an overall task, but are themselves goal directed (and hence subtasks) only in the trivial sense that the end-state of the element's activity is considered its goal. The definition of "parallel process" was given in terms of tasks and subtasks, rather than merely in terms of concurrency of activity, for the reason that we are exploring the possibility of conscious parallelism within an individual mind. If it is to be a conscious parallelism, there must be consciousness of each of the concurrent thought processes. Consider the overall task of resolution of a mathematical problem. To the extent that the problem admits of parallel division of labor, each strand of concurrent consciousness needs be occupied on a subproblem.[1] For example, consider the addition of five and eleven, the multiplication of four and two, and the division of the former by the latter. Assume that the addition and multiplication can be consciously performed in parallel, with the results grouped for the division operation (and these operations might themselves be subdivided: if, for example, they were performed in binary arithmetic—with calculation rather than rote response—at least several operations would be necessary). On the assumption that parallel consciousness is possible, the question is the limits of subdivision of an overall problem into subproblems. Regarding the sample problem [(5+11)/(4x2)], the question arises whether there are recognizable limits beyond which no task is represented. It might be maintained that the mere holding of separate integers in conscious parallel awareness would itself be an instance of parallel consciousness. Doing such by itself, however, is arguably not part of the dynamic of solving this particular problem, and therefore is not to be considered a subtask of the overall task. To the extent that it is debatable whether or not holding the various numbers of the sample task in parallel consciousness is itself a subtask of the overall task, the notion of subtask is not clearly delimited. There may yet be an intuition that there is a point, however indistinct, below which no task is represented.

Problem-solving procedural thought is not the only sort of mentation in which the question of the possibility of parallel consciousness arises. Typically, one sees two adjacently parked automobiles at a near distance. One is rapidly aware of the twoness and adjacency, and perhaps in most circumstances less immediately aware of the brands of the two automobiles. As an introspective experiment, one ought to try to hold the brand names (say, Chevrolet and Honda) in parallel conscious awareness. In my own case, I discover an inability to do this simple task. There is no problem solving here and no "movement" of thought other than consciously to bring forth a label (I have a similar problem simultaneously holding in consciousness the words "twoness" and "adjacency"). Nevertheless, there is an ostensibly implicit parallelism in the seeing of more than one parked car simultaneously. Such seeming parallelism in perceptual consciousness will be considered further below.

It might be argued that in various manners parallel, consciously directed thought is an established fact. If the historically inherited thought of society is a whole, of which the thoughts of individuals (influenced by, and influencing the whole) are parts, then the focus on the intellectual capacities of individuals might strike some as symptomatic of a deplorable atomism. Given a plurality of minds concurrently engaged in related tasks, and occasionally sharing their thoughts, a parallelism of sorts emerges. With regard to individual consciousness, the sort of inspired thought that arises apparently *ex nihilo* as the solution to a problem might indicate an unconscious-to-conscious parallelism. What is speculatively envisioned here is that such discontinuity in serial consciousness is the result of concurrent unconscious operations "tracking" a conscious flow of thought. Yet another form of ostensible parallelism within the individual is indicated by the following two examples. In being attentively conscious of a plurality of items in one's visual field, one often has little difficulty tracking the simultaneous motions of two or more entities, even if the various motions are not uniformly translatable. Similarly, in being aurally attentive one often tracks the simultaneous separate voices of ensembles such as string quartets.

Consideration of parallelism in our collective thinking serves to loosen fixation on the individual mind. It nevertheless remains that our serially-proceeding consciousness is a bottleneck in the production of thought within the individual. One need only reflect on one's experiences concerning various matters of intellection; for example, the effort in making a contribution to a field of thought. There is the effort to acquire further understanding of the field, the initial construction and subsequent revision of one's work, and so forth. Noting that others are simultaneously engaged in similar work, and in communication, does not remove the frustration one occasionally experiences by reason of one's slowness to comprehend. In part, it is our serially-based conscious thought that extracts a price in time.

It must be conceded that concomitant brain activity accompanies serial consciousness, forming a sort of parallelism of the "unconscious" to the conscious, however problematic the causal relations between these two seemingly different sorts of activities. By hypothesis this is not a parallelism of activity within consciousness. Aside from the possibility of forming habits which promote sudden insights, we manifest no control over the process. This lack of control contrasts with our conscious ability to intervene and redirect our consciousness. It is noteworthy that as the sudden insight (or random thought, or wayward but associated thought) appears in consciousness, it takes a position in the serial flow thereof, rather than occurring concurrently with another flow of conscious thought in the individual.

Regarding the apparent parallelism in noting simultaneous events, this ability has been briefly touched on above in the example of the simultaneous presence of two automobiles in one's field of vision (accepting that the relatively static presence of perceived motionless objects is an event). The parallel or serial nature of such perception seems borderline, or vague, in that one is conscious of separate presences or motions in one's visual field, but one is perhaps incapable of concurrent focus on these separate entities or events. One observes an automobile approaching a pedestrian at an intersection. One is aware of the respective motions of the automobile and the pedestrian, in that one is simultaneously tracking these motions. It remains uncertain, however, that such awareness manifests a genuine parallelism. An apparent parallel attentiveness to independent motion may be a shifting from one motion to the other; that is, a serial activity. Various time-sharing computation gives the illusion of parallel processing in such a manner, although processing is proceeding serially. One's attention might be drawn to an accident-in-the-making, with subsequent rapid darts of attention from the automobile to the pedestrian. In more ordinary circumstances, such as a casual approach to a street crossing, one likely pays little attention to automobiles and pedestrians, except for those immediately presented. Although most of the automobiles and pedestrians are virtually unnoticed, those that are in one's visual field are the objects of a perceptual sub-awareness, and are apt to be brought in the forefront of attention by unexpected "triggers."

The preceding discussion omitted a number of subtleties. The individuation of motions is a more complicated affair than was suggested. The moving pedestrian was considered as a single instance of motion. However, the pedestrian could have been considered an assemblage of motions. The motions of the pedestrian's arms, legs, and torso are separable. If the moving pedestrian is roughly centered in the observer's visual field, then the separable motions of various pedestrian parts are obviously present in the observer's visual field. As in such cases as the unnoticed pedestrians and automobiles, the various bodily motions of a single pedestrian are arguably in the observer's sub-awareness, and if brought to full consciousness would likely be attended serially. A plausible suggestion is that motions within a visual field are individuated in consciousness either by exceeding a threshold angle of separation or, more generally, by varying focus of attention.

Remarks regarding attending independent motions in an observer's visual field similarly apply to awareness of individuated sounds in one's auditory field. Consider again the example of attending to the separate voices of a string quartet. There is the possibility of perceiving the separate voices of the quartet as a blended whole by a relaxation of attention. Such blending via relaxation is similar to the visual situation of a pedestrian crossing an intersection, attending only to objects of immediate relevance while somehow "blending" the remaining visual phenomena. Further, when the perceived motions are intuitively translatable, as for example a flock of birds flying in close formation, a blending of the plurality of moving objects into a unity is apparent. When such blending occurs, the conscious attention to the blended whole is serial.

Why is the standard serial consciousness not all that humans might require? The answer is that there are various significant and lengthy cognitive tasks which are divisible into lengthy subtasks which do not require sequential thought. For such tasks, a serial processing is arguably not as satisfactory as a parallel processing. But if there is no pressing need for the sort of improvement that parallelism apparently offers, then the examination of the possibility of such parallelism appears useless speculation, except for such insights as it might offer into the nature of our actual thought processes. It should be emphasized, however, that there is more use for parallel thought than the comparatively trivial mathematical example given above. If humans had the ability simultaneously to hold large amounts of complex theories in consciousness, for example, the possibilities of insight might be greatly enhanced. An example is suggested by current software technology. "Windowing" user interfaces have improved productivity by allowing several tasks to be concurrently displayed. Still, one's conscious focus moves serially among the windowed operations. Imagine the capability of one's very consciousness to "window"; that is, to represent several thoughtful endeavors to simultaneous awareness.

On the assumption that there are benefits to an individual's having a vigorous parallel consciousness, the question of the possibility of such consciousness merits further focus. Such a question has two sides: inquiry into reasons against such possibility, and inquiry into how such a possibility might be realized—provided results of the former inquiry do not stifle the further inquiry. As to the former inquiry, reasons for limitations on human abilities range from logical or mathematical impossibility, through factual impossibility (in the sense that the manifestation would conflict with well-entrenched law or theory), to the so-difficult-as-not-to-be-done-yet (for example, attending to a large group of speakers simultaneously presenting papers on unrelated topics).

According to some writers of fancy, such as Robert Louis Stevenson, there is little logical difficulty in the conception of one body shared by two persons (although perhaps not simultaneously). There are factual examples of personality disorder in which such sharing is supposedly the case. To speak of persons, rather than bodies, how intimately is a person's numerical uniqueness tied to a single flow of consciousness? Comparatively durable, simple bodies, having little material exchange with their environment, are typically thought to be grounded in their numerical uniqueness by spatiotemporal continuity. Material configurations subject to higher-level characterizations, including those of social or legal sort, and/or in significant material exchange with their environment complicate, if they do not invalidate, the "trace" criterion of identity over time. Allowing the familiar discussions of material object identity over time as paradigmatic, or minimally as suggestive, the question becomes to what extent is a person's numerical uniqueness tied to a single flow of consciousness? As a logical thesis, namely that personal uniqueness necessitates a single path of consciousness, the thesis is doubtful. If continuity of thought is a necessary condition for the individuation of persons, most individual persons would no longer be regarded as such, unless the continuity requirement were specified so loosely as to be meaningless (consider, for example, the degree of continuity of thought in Molly Bloom's famous soliloquy in Joyce's *Ulysses*).

Extending individual parallel consciousness to the limit might be logically impossible. Consider the supposition of an individual having nothing but parallel consciousness. Waiving the difficulties in detailing this supposition, a general consideration of it strains intuitions of consistency. The individuality of a person apparently requires a single serial flow of consciousness having executive and coordinating functions such that parallel conscious flows are to some extent integrated with it. But this supposition of radical individual parallel consciousness is an extreme case. An electrical layout having separate current flows with no common point of origin or point of junction would not be considered one parallel circuit, but rather two or more independent circuits. Still, such intuitions and analogies are less than conclusive. Unlike non-interacting electrical circuits, strands of consciousness are tied to a single physical person, considered as a continuant over time. Further, if such a continuant were to have only parallel conscious flows, if one such flow were in some manner performing integrative and executive functions, the intuitive difficulties of such an extreme situation would be mitigated.

Claims of the impossibility of parallel consciousness supported by putative physical or psychological facts should be approached warily. People in radically different cultures have on occasion exhibited reported abilities which have astonished Western observers. Among notable examples are those of certain Eastern holy men to control such physiological aspects as their blood flow to an extent thought to be outside the range of willful control.

The question arises as to the means by which it would be established that some people have genuine parallel consciousness of the sort here speculated upon, rather than the sort arising from "split-brain" physiological commisure of the corpus callosum. Although this latter sort of parallelism is readily encountered in the philosophical literature, it is not the sort of parallelism contemplated here. Rather, what is of concern here is a sort of parallelism under some form of conscious, executive control. Given the privacy of consciousness in practice if not in principle, introspection is likely to be involved in establishing the existence of such parallelism. Granting honesty of reports, the difficulty remains for the introspector to distinguish between genuine parallelism and fast serial switching from one flow of thought to another. If certain intellectual tasks having indivisible subparts of sufficient size can be specified, then perhaps a test that avoids this difficulty can be devised.

It is a reasonable supposition that any factual difficulties in achieving parallel consciousness are due to factors in human neurophysiological constitution and/or human consciousness itself. There is of course the large question of the existence and relation of consciousness to neurophysiological activity, which question had been set aside early in this study. There is the further question as to the extent sociohistorical considerations intrude on the possibility of parallel consciousness. Even if consciousness reduces to, or is eliminated in favor of, neurophysiological activity, such activity is situated in cultural context. Recall the above point regarding possible differences of abilities in culturally distinct populations. Accounting for such differences, if they prove genuine and germane, might well involve more than appeals to immediate consciousness or to neurophysiological happenings.

In a speculative spirit, the following considerations are offered as suggested factual possibilities accounting for the difficulty in achieving parallel consciousness. Typically, a flow of conscious thought presupposes a background from which the objects in the foreground of consciousness derive much of their significance. For example, in consciously thinking about one's automobile, whether with a specific purpose in mind (e.g., waxing it) or engaged in (simple?) reverie, one possesses background notions such as automobile and ownership, and if one were to foreground these notions it would be against the background of yet other notions. On the assumption that such background is large in comparison to the conscious foreground; that is, that the background involves an appreciable amount of memory, it is possible that there is insufficient remaining background capacity for supporting two or more concurrent lines of conscious thought. Or, to speak the language of networks, it may be difficult for the various neural networks supporting the two or more conscious flows to activate simultaneously. Such suggestions are questionable on the ground that various conceivable concurrences would draw upon the same background or the same neural net. However, these suggestions lead to a further possibility regarding the background system activated in the support of consciousness. If the mind or brain has something akin to an operating system, suppose the system is programmed or even hardwired to match one background to one line of consciousness in the following manner. On initiation of the line of consciousness, the mind or brain closely monitors the flow and activates appropriate related background as the flow progresses. If different lines of concurrent consciousness, however qualitatively similar, demand different related backgrounds, it might be difficult for the operating system to support a synchronization of different independent communications from the background to the foreground of consciousness. The overall system might block attempts at concurrency undertaken by the conscious agent (such an agent being that part of consciousness exercising the control functions ordinarily ascribed to agency).

Despite mention of the word "mind," the preceding speculations are from a computational-neurological perspective. The following suggestion is offered from a psychological standpoint. Assume that a flow of consciousness has a certain flavor of personality associated with it. Although afflicted with vagueness, the notion of personality has been employed by various philosophical and psychological traditions. At least some of the vagueness of the term "personality" (and related terms such as "character") is due to the lack of a clear counting principle by which to differentiate personalities. Suppose, nevertheless, that different flows of consciousness in some manner have different associated personalities. This supposition involves stretching intuitions of the term to the extent that different episodes of consciousness (waiving difficulties of specifying such difference) of what would normally be considered a single person entail different personalities. Further suppose that it requires only small difference in the content of consciousness to determine a different personality. Sameness of personality is entailed by some sort of principle of continuity of consciousness. An overall person's "master personality" would be regarded as a unique summation, and possible generalization, of such conscious flows personalities. A sort of preservation-of-the-immediate-self function might exist to prevent more than one consciousness-personality stream at any given time. If so, those with certain psychological disorders might achieve concurrent consciousness with less difficulty, provided they could be motivated to attempt it. Also, those who have most thoroughly eliminated "personality" from their consciousness might be in a favored position.

There is much philosophical thought regarding the relations among the computational, the neurological, and the psychological. Additionally, the field of evolutionary biology is often brought into the discussion. Our characteristics, our behavior, are often ascribed as a partial function of adaptation to environmental constraints. To the extent that such ascription is valid, the possibility lurks that human ancestral survivability was aided by, if it did not necessitate, a focus of consciousness that prohibited parallelism.

Given that there is much unknown about the matters under discussion, much that science has yet to discover, the following is offered in a continuing speculative spirit. One manner in which parallel consciousness can be approximated is by the social pooling of the standard serial consciousness of individuals. The value of this pooling is at times underestimated by those who have never engaged in a large collaborative project or in collective brainstorming sessions. Another manner in which some of the presumed power of conscious parallel thought can be realized is through a kind of training. Given that non-conscious "automatic" thought is at times apparently parallel, in familiarizing oneself with a realm of thought to the extent that some of the thought processes therein become automatic, then at least some thinking which formerly demanded more lengthy serial consciousness can be done with greater dispatch. As an example, consider the manner in which an experienced mathematician might take in various computations "at a glance" while the novice must work through them one after another.

The preceding two suggestions do not involve actual conscious parallel thinking. Assuming that such thinking is possible, it may simply be a matter of engaging in certain forms of mental exercise to develop such mindfulness. Perhaps a certain amount of neurosurgery is necessary, or that a form of implanting, biological or nonbiological, will realize such consciousness. Various presently available and/or future pharmaceuticals may allow access to this mode of thinking.

Extending human capabilities in these manners is not without risks. Fundamental changes in the general way humans think have to this point been limited to the sociohistorical realm. Cultural relativity aside, the differences between contemporary thought and the thought of, say, Classical Greece, mainly regard scientific and technological matters. Not that such a difference remains confined to these areas, rather than permeating other realms of thought. But there is otherwise little reason to believe Plato or Aristotle thought less acutely than any contemporary exemplars. Assuming the majority of humanity could substantially improve their "raw intellectual abilities" in the manner under consideration, or in other speculative fashions, what would become of us? One way or another such possibility may soon be upon us. This is arguably worthy of consideration.

NOTE

1. Perhaps there are cases in which such goal direction is absent (for example, repetitively humming a melody in consciousness while solving a problem). Such cases can be overlooked, however, for the reasons that they are less significant than task-oriented cases and that they admit of a similar analysis.

Chapter Six

Mindful Seeing

But how shall he support this enthusiasm itself? The bent of his mind relaxes and cannot be recalled at pleasure; avocations lead him astray; misfortunes attack him unawares; and the *philosopher* sinks, by degrees, into the *plebeian*. —David Hume, *Dialogues*

Two tables! Yes; there are duplicates of every object about me—two tables, two chairs, two pens. —Arthur Eddington, *The Nature of the Physical World*

Philosophers are generally aware of what C. P. Snow termed "the two cultures"; the culture of natural science and that of the humanities. Some philosophers are occupied with questions related to Snow's concerns. Many philosophers inhabit one or the other of these two possibilities.

Continental philosophers, for example, are typically regarded as humanistic in their subject matter and methods. Analytic philosophers, such as those inheriting the concerns, if not the methods, of the logical positivists, at times appear either scientific, or aspiring to be scientific, at least in the loose, intuitive sense of the term.

Regardless of one's philosophical outlook, philosophers and the vast majority of humanity live in the world of the humanist. This is the world of colors, smells, tastes, feels. It is also the world of tables, chairs, rocks, rivers. It is the world of appetites and emotions, and human relationships. It is not the world of mathematical entities, idealizations of physics, chemical elements, DNA.

Allowing the simplification that *two* "worlds" are indicated here, their distinctness is in question. They are less than distinct in various ways. The boundary between them is unclear in many cases. In those instances with clear boundaries, it may be unclear where the human participant's awareness lies. Further, if there is clarity as to boundaries and as to an individual's awareness, what may be less than determinate is the shifting of an individual's awareness from one state of mind to another. More could be added by way of challenging the lucidity and viability of this notion of two worlds. Yet there is something to what Snow was urging, and something to Eddington's two tables (one of ordinary experience, the other of microphysics). If the boundaries between the worlds are somewhat indistinct, and if the consideration of them as two worlds, rather than several or many or somehow one, is in question, such will do little to alter the main line of discussion to follow.

There remains the philosophical difficulty regarding the distinction between the world, or worlds, and our experience of such. Is there a plurality of worlds, or simply a plurality of experiences? While some consider such a question illegitimate,[1] no position need be taken on this matter here. The concern here is with experiences. How the world is, or if it is, apart from human experience, are questions that can be left unaddressed.

In brief, what will be considered here is the possibility that humans transform themselves from beings largely having humanistic experiences to humans largely having scientific experiences. This may seem to be no more than Eddington and Snow are urging, to the extent that they favor a scientific understanding over a common one. Neither Eddington nor Snow, however, suggest that we lose the common standpoint from which most of our experience is determined. Rather, to the extent that they are offering a similar message on the topic, they indicate a need for a greater scientific understanding of matters that do not yield to common understanding. They do not presume that people, themselves likely included, will cease to spend most of their lives in the world of heretofore common experience.

The thought of spending most of one's life experiencing the world in a "scientific frame of mind," for example considering everyday objects as physically interacting mass points, is sufficiently alien that it might be believed a suggestion no serious philosopher has made. However, a similar suggestion has been made by at least one classical philosopher, and a contemporary philosopher has broadly indicated such a possibility.

In the *Phaedo* Plato is perhaps at his most extreme on this point, promulgating the view ascribed to the Pythagoreans that the body, with its senses and appetites, is the prison of the soul from which the soul must depart if it is to enjoy an unfettered existence (65e–67a). Given that Plato is speaking of a disembodied life, his remarks are not strictly to the point of concern. Plato does speak here, however, about life as a preparation for death (defined as separation of body and soul), suggesting that as much as possible, one should live one's life in a "bodiless" frame of mind.

It should also be noted that the Platonic forms, perhaps making their first appearance in Plato's dialogues in the *Phaedo*, are not completely coincident with scientific concepts. Considering that some of the forms, such as beauty and justice, regard matters of value, many would object to the ascription of a scientific outlook to Plato on the basis of his theory of forms. Nevertheless, Plato sought a rigorous understanding of these ostensible entities. He clearly thought that something like a theoretical science of them was possible or, put another way, he thought that such an understanding of them was in some manner foundational for the sort of knowledge later to be termed "scientific." That Plato's understanding of natural phenomena was limited by his historical circumstances, and that he arguably gives insufficient place to mathematics in his overall scheme of understanding (although he by no means ignores mathematics), should not detract from the similarity between the notion of a scientific outlook, as such is understood now, and Plato's vision of a life in pursuit of the forms.

Paul Churchland occasionally evinces a visionary outlook. In his noteworthy essay, "Eliminative Materialism and the Propositional Attitudes," Churchland speculates about nonsentential, neurally-based modes of communication. Such visionary speculation, coupled with his stated belief (in *Scientific Realism and the Plasticity of Mind*) that humanity will generally come to adopt a physiological vocabulary which replaces commonsense psychological terminology, suggests the above viewpoint of experiencing the world in a scientific frame of mind. Granted, in the latter work Churchland does not consider the rough percentage of waking life spent in a scientific frame of mind, and in the former work Churchland is speculating on radically different modes of thought than those anyone has heretofore engaged. Still, there is recognizable overlap between Churchland's flights and what is under consideration.

Between Plato and Churchland, both chronologically and philosophically (at least with regard to the investigated topic) lies Descartes. Descartes' dualism is a match to Plato's, and his rationalistic disavowal of knowledge through the senses is the equal of Plato's. However, Plato's claim in the *Phaedo* that proper knowledge is an otherworldly experience necessitating death is not echoed by Descartes (and, as is well known, in other works Plato does not insist upon a separation of mind and body as prerequisite for pure knowledge). Nor does Descartes share Churchland's vision of a sub-sentential form of knowledge representation, a vision that likely never occurred to him. Further, Descartes gives the impression of being more relaxed than Plato, at least the Plato of the *Phaedo*. One need not "practice death" to engage in pure understanding. It is sufficient that one has freed one's mind "of all kinds of cares"; that one is "disturbed by no passions" (*Meditations*, first meditation). While so meditating, Descartes recommends obeying the laws and customs of surrounding society (*Discourse*, part three) and in general doing all possible to avoid external and internal impediments to thought.

There is scarcely an academic philosopher who has not read and taught the *Meditations*. The work serves as an introduction to concerns in areas of epistemology, metaphysics, philosophy of mind, and philosophy of religion. It is also safe to say that among English-speaking philosophers, if not others, the work is not often pursued as a commentary regarding philosophical lifestyle, despite Descartes' various remarks to this purpose, such as the following:

> It is not enough to have made these observations; it is also necessary that I should take care to bear them in mind. For these customary and longstanding beliefs will frequently recur in my thoughts . . .[2]

Such remarks, which occur with surprising frequency throughout the work, are easily passed over as dramatic literary flourishes inserted to delight, rather than edify, the reader. They may be that. Alternatively, Descartes might be linked with both Plato and Churchland, among others, seeking a different mode of life on earth than that of the commonplace, whether such commonplace life be of high or low culture (but including the occasional foray into theoretical thought). Descartes might be recommending a general residence in the world of theoretical thought, with only occasional forays into the commonplace.

One more example: Although often popularized as the head-in-the-clouds scientist, Albert Einstein was a person with earthly characteristics, particularly as a younger man. He occasionally struggled against his mundane side, however, in the interest of his theoretical thought. As various appetites abated in older age, Einstein came more to resemble his popular image. Shunning the public in later years at Princeton, he was known to take walks accompanied by people assigned to monitor his paths, lest he step into Thales' well while muttering "I must think."

A significant possible difference between Plato—at least the Plato of the *Phaedo*—and Descartes should be emphasized. Descartes speaks of the life of mind, to be lived while one is ensconced in the commonplace world. This mundane world must be accorded consideration, according to Descartes. One must obey its laws and customs, and in general act as a properly socialized individual. Plato speaks of a departure from the commonplace world. The departure, being death, is a radical departure. There is the suggestion, however, that death is the subject because it is Socrates' last night on earth. The practice of death suggested for the living might be other than Descartes' meditative retirement. Renouncing the body's inclinations is a call for an intellectual asceticism of a sort the sophisticated Descartes could not fully adopt for himself, but could approve of as consistent with his recommendation. There is, however, an alternative approach to living consistent with Plato's goal of renouncing the mundane, although inconsistent with his suggested means. Consider an amended practice of death as a program for the living philosopher to engage the world theoretically, rather than commonly. In this practice, one's sense perception would be at some level the same as that of the dweller in the commonplace, but one no longer sees rocks and trees, tables and chairs, because one's ordinary experience is charged with a different conceptual scheme than that of the commonplace.

The above assumed difference between an amended Plato and Descartes places the amended Plato, and Paul Churchland in his visionary moments, on one side of the divide, Einstein and Descartes on the other. These thinkers are not divided in their allegiance to a scientific, proto-scientific, or quasi-scientific outlook. Hence, prior to considering the radical alternative posed by the amended Plato/Paul Churchland style of living, the general question is raised as to the value of the outlook shared by Plato, Churchland, Descartes, and Einstein, in the face of competing outlooks. Other questions that can be directed equally at them include those of proper time in life, and those of resources. Are we to wait until age dampens our earthly desires before spending most of our waking hours in the realm of thought? When we do so choose, will we need human resources such as attended Einstein? Are there sufficient resources for all to embark on this course, or only a privileged few? Further questions are directed at the more radical proposal ascribed to Plato and Churchland. What is it to charge one's sense perception with a different conceptual scheme than the ordinary? How would such be accomplished? For what reasons should it be done? By the end of this essay, answers to these questions will hopefully be given, sketched, or indicated.

In the philosophical literature there is the occasional call for, or noting of, alteration of ordinary perceptual experience. Bergson comes to mind here.[3] The writings of Benjamin Whorf suggest that radically different language-users see the world differently.[4] Among the most emphatic points of post-positivistic philosophy of science is the generally shared thesis that perception is intermixed with whatever the current conceptual scheme happens to be, so that a sharp theoretical/observational distinction is untenable.

The writings of Carlos Castaneda supply yet another example of seeing the world differently. In these writings, Carlos apprentices himself to a Yaqui "sorcerer," Don Juan. Although these writings may be largely fiction, there are noteworthy passages such as the following from *A Separate Reality*:

> "Then, Don Juan, you don't see the world in the usual way anymore."
>
> "I see both ways. When I want to *look* at the world I see it the way you do. When I want to *see* it I look at it the way I know and I perceive it in a different way."[5]

People, for example, appear to Don Juan as

> "fibers, like white cobwebs. Very fine threads that circulate from the head to the navel. Thus a man looks like an egg of circulating fibers."[6]

It should be noted that Don Juan does not require hallucinogens to see people in this manner, although he recommends such to Carlos as a first step in such altered perception.

To pursue discussion of such examples either in the academic literature (for example, William James' *Varieties of Religious Experience*) or the more popular literature of altered states of consciousness, would be to move from the topic of concern. The focus is on experiencing the world in what would currently be termed a scientific frame of mind. To repeat, many people have spent significant parts of their lives in a theorizing state of mind. But to do so, they have typically avoided the world of ordinary sense experience while in this state of mind. At the least, the possibly fictitious Don Juan manages to bring his theoretical outlook into his daily perceptual experience.

Many were attracted to Castaneda's writings at the time of their publication. The current inclination is to dismiss the topic of altered states of consciousness. Perhaps such dismissal is an intellectual backlash against the earlier enthusiasm. In any case, this inclination is checked by the observation that perceiving the world in a scientific frame of mind, rather than a commonplace frame of mind (i.e., "commonplace" relative to whatever the current state of scientific inquiry) is arguably to be in an altered state of consciousness. Why preference one such state over another?

The answer to this question is in principle the same as the answer to the question of the value of various forms of life: the aesthetic, the sensual, the acquisitive, the philosophical, and so forth. Certain forms of life can be universalized in the sense that projecting them onto humanity allows for humanity's survival, while other forms do not. Through science, mediated by technology, humanity has survived, and increasing segments of the overall population have prospered. The urge to bring scientific consciousness into everyday life allows for a fundamentally less epistemically alienated life than other suggested forms of life. This is not to disallow aestheticizing ordinary perceptual situations, or to restrict the aesthetic experience to the perspective of social realism. It is a commonplace of the cultured life that one occasionally aestheticizes one's ordinary perceptual circumstances, conditions permitting. Rather, following Plato, among others, it is to place a higher value on cognitive experience than that of the aesthetic, appetitive, acquisitive, and so on.

One might grant the value of scientific thought, while maintaining that ordinary perceptual experience is best relegated to the commonplace. By analogy, one might appreciate the grounding of various "higher-level" scientific outlooks in the physics of microparticles, while maintaining that the scientific study of, say, volcanic activity, is best done at the macrolevel, with appeal to the microlevel when appropriate. Quine, James, and others have praised the implicit theoretical framework of commonsense-object talk. Why look at the ordinary world in any other way, unless special circumstances so dictate?

The notion of ordinary, or commonplace, perceptual experience is wide. Among the possible discriminations of this broad area, the following two are pertinent to the pursued theme. The realm of medium-sized objects, generally well-behaved continuants in a spatiotemporal field, is typically considered the world of commonplace perceptual experience. Philosophers at times emphasize the less-than-given nature of this realm, and the manner in which human cognition organizes this realm from the welter of sensory stimulations. For some philosophical purposes, there is little difference between this realm constituting the experience of a paleolithic person as against that of the experience of a contemporary person. There is, however, a more detailed understanding of the realm of medium-sized objects, according to which a person's commonplace experience of this world is shaped by aspects of prevailing theory. Such experience embraces light switches (and not merely small brown objects of limited degree of motion), heating vents, diesel-powered trucks, and so on.

Philosophers generally agree that ordinary perception is charged with significance extending beyond such comparatively simple properties as color, texture, and shape. What is at issue here is the proper degree of such significance. What should people be held to see? The answer to this question, if an acceptable answer is possible, will obviously be general. One person might see a truck while another sees a diesel-powered truck. Other things equal, is the latter's perception preferable to that of the former? To the extent this example allows, yes, because the latter's perception is more cognitively detailed, and this is a virtue, other things being equal. Of course, extended consideration might result in a retraction of this judgment due to a violation of the ceteris clause. If the latter person is knowledgeable about trucks, but otherwise sees the world in more restricted manner than the former person who is, say, an accomplished biologist, then the former's overall history of perceptual experience is likely to have more value, measured by cognitive standards.

The cognitive value of one's perceptual experience is not determined solely by depth and detail of object identification. The extent to which the perceiver perceives interconnections among the objects, typically dynamical interconnections, adds cognitive value to the experience. Ultimately, there are too many sorts of enhancements to list here, but such general examples are indicative of the point. To consider one more actual figure, there is much anecdotal evidence that John von Neumann, a person who lived very much in-the-world, enriched his perceptual experience by occasionally "mathematicizing" it in various manners according to physical principles. One might respond that we are most of us not von Neumanns. This response will be considered below.

Evaluating perceptual experience raises familiar problems. Notable in this context is the problem of competing modes of perception. One person perceives in what might be termed a mundane fashion. Another aestheticizes their experience. Yet another perceives human encounters in social-scientific terms: as an economist, or a psychologist, or a sociologist, or a historian. On the assumption that various of these perceptual modes cannot be had simultaneously, how is their respective merit determined? This question has been partially addressed above in the discussion following the Castaneda example. Regarding the wider range of possibilities under consideration, the question is to some extent a reflection of the question of the value of a certain style of life. The latter question, of course, is at the core of much of the current and traditional ethical deliberation in philosophy.

With proper regard for the magnitude of the latter question, the following can be said. Assume that a life of development and exercise of one's cognitive powers is preferable to those forms of life not emphasizing such. On that assumption there is a prima facie case for cognitively enhanced perceptual experience. Such experience need not be enhanced by natural scientific considerations. In one's dealing with others, including perceptual experience of others, one might enhance such experience by fitting it into a framework of social scientific theory. Those who "psychologize" others as they interact with them are doing this. Many are willing to accept this sort of enrichment of that part of daily perceptual experience having to do with personal interaction. Fewer are likely to consider enriching their perceptual experience by injecting natural-scientific matters into the flow of the experience.

Many people occasionally aestheticize their perceptual experience. In keeping with the Platonic judgment that the aesthetic, unless informed by the cognitive, is not the highest form of experience, and should be de-emphasized in favor of the cognitive, such experience is not recommended here. Yet if one were to choose between the mundane and the aesthetic, the only reason for choosing the former would be that supplied by Plato; the aesthetic, in the form of poetry, drama, and painting, draws one away from the cognitive, or misleads one into conflating the aesthetic with the cognitive. One who believes that people cannot live by bread and science alone will reject this wholesale devaluation of aestheticizing perceptual experience.

Indeed, the idea of charging one's perceptual experience with scientific thought seems wayward in several respects. Engaging in difficult thought often requires that one turn away from perception in the interest of concentration. Perception is typically joined with real-time being-in-the-world. Crossing a busy street, for example, is not the time for theorizing about one's perception. One's dealing with others often has an immediacy to it that prohibits flights of thought. And, as was noted above in connection with the example of von Neumann, most people are satisfied to take their perceptual experience on its own mundane terms, and would find the suggestion that they imbue such experience with scientific thought burdensome, if not inconceivable.

The force of these objections is undeniable. Worldly concerns coerce people to attend to the here-and-now of their perceptions. Most people take perceptual experience as a relaxing alternative to thought, the latter of which they either avoid or perform as an obligation to be discharged as quickly as possible. In this respect, the life of mind resembles the life of virtue; many praise it and few live it.

Taking human nature in some general sense, and human life as currently lived in developed societies (it is assumed that most people living in underdeveloped societies lack various necessary conditions for scientific thought), the above objections cannot be overcome. Still, one might press the point by questioning the two givens of human nature and normal human life. It was remarked above that we are not all von Neumanns. But why aren't we? To date, we have only an "outward" answer to this; most of us do not behave as von Neumann. Accepting that to some extent we do not behave as von Neumann due to differences in our nurturing, there is the thought that a partial reason we do not so behave has to do with variances in our neurophysiological makeup. This prompts the question as to what our societal, or perhaps species-wide, policy should be if we ever acquire the means to make ourselves neurophysiologically similar in relevant respects to von Neumann.

If we had the means to make ourselves into beings with intellectual powers similar to those of a von Neumann, ought we do so? We currently immunize against various scourges on a species-wide scale, and many of us accept the principle that all people have a right to literacy and a right to as educated a status as resources allow. If we had the knowledge and means to improve our intellects to the extent broadly indicated—that we acquire intellects similar to von Neumann's—it would seem initially that such acquisition should be mandated for all.

Granting the large assumption that all people could be given superior intellects, perhaps at or prior to birth, questions arise as to the sort of human life that would result. Who does the repetitive, laborious tasks variously necessitated by our modes of production and distribution of goods and services? How do we care for ourselves as a society of thinkers? What becomes of life bereft of the small pleasures for which we no longer have regard?

If humanity transforms itself into a world of beings with the intellectual talents and interests of someone such as von Neumann, large social, economic, and political transformations will have been made. As we live today, with unequal distribution of goods and services, various prerequisites for such global transformation are lacking. Perhaps the transformation could be done on a smaller scale (that is, with fewer people), such people then serving as a vanguard for sweeping changes necessary for universal intellectual enfranchisement. By whatever specific means, if humanity undergoes such wholesale profound transformation, the background conditions of life will have changed. It can be expected that muscle labor will be replaced by machine labor, that such machines will either be self-servicing or will be repaired by other machines. It can be expected that repetitive intellectual labor will be done by "smart" machines.

The devil is in the details. It should not be expected that details be provided here. In keeping with the point of this chapter, the focus in the remainder of this study will be on the ostensible duality of theoretical thought and perception.

In the *Phaedo* Plato sets the tone that has echoed for over two thousand years. Sense perception is in opposition to theoretical thought. Note that the opposition here is of at least the following two related sorts: epistemologically, sense perception and theoretical thought (perhaps qualified as "rational" thought, in that some empiricists would argue that theoretical thought is itself an empirical matter) are opposed as to the sort or level of knowledge they supply. As a mode of awareness, they are opposed in that one may largely live the life of sense perception or the life of thought. While some philosophers, notably Kant, have sought to reconcile the first sort of opposition, it is a difficult to find a Western philosopher attempting to reconcile the apparent duality of thought and perception as modes of living.

This last observation returns us to the starting point; the profound alienation of a large aspect of human life. We live most of our lives in the realm of the senses. Yet we take the realm of thought to be a higher realm in which to dwell, although difficult for many of us to dwell there. What is almost universally accepted is that we cannot dwell simultaneously in both realms. The response to the example of von Neumann—most of us lack his abilities—has been considered. What remains to be considered is the claim that even one such as von Neumann did his most powerful theoretical thinking secluded from his perceptions. Rather, his perceptions, when he so chose, were infused with applications of theoretical thought simply for his amusement.

As was stated above, the world would need be a profoundly different place for humanity as a whole to possess and exercise intellectual capacities similar to those of von Neumann. Whether such advancement is possible, it must be conceded that at that stage people would be excluding perception from their deeper theoretical excursions. Still, at that stage there would be the enjoyment of perception in a manner in which perception is rarely enjoyed now. If this seems a comparative trifle, something more is conceivable. Namely, that humanity reaches a stage of development in which people theorize extensively while interacting perceptually with their environment.

However much extreme rationalists have denigrated sense experience, the world of sense experience is the world inhabited by our embodied selves. It is some measure of Plato's genius that in the *Symposium* he noted that pure thought needs an inspirational component, a driving force, an eros, to put it in motion. He believed this driving force was initiated by sensual elements in perception. Such driving force could then be transformed from a response to the sensual to a response to the theoretical. Viewed in this manner, the *Symposium* modifies the radical "practice of death" of the *Phaedo* by acknowledging the role of sensually-charged perception in putting one on the path of higher knowledge. To some extent, the movement of the *Symposium* is reversible in the following manner. While the theoretician of idealizations takes delight in the abstract structures of thought themselves, much theorizing can be done with the aim of application to the perceived world. Desire for application is often the "eros" of theorizing, separate from the "eros" of theorizing for its own sake.

Imagine a world in which people are largely motivated to pursue theoretical interests, and in which they have the cognitive ability to pursue these interests as they interact with the perceived world. These people obviously possess internal cognitive resources not presently available to most, if any, of us. There is additionally an overhead of external conditions necessary for maintenance of goods and services supporting such lives, and people will be generally aware of this overhead in their various theorizing regarding their perceptions. Which is but to say that people in this situation will be living well, aware that they are living well, aware of the requisite conditions for living well, and aware of their roles in sustaining and furthering such well-being.

Speculations of future possibilities are typically fulfilled as the inverse of their detail. Examples of what is being alluded to are nonetheless desirable. The following are offered in the spirit of suggestion. These offerings are restricted to the modality of vision. Any observant owner of a canine is aware of the olfactory dimension from which humans are largely excluded. Granted, what is being considered is less a heightening of perceptual experience than the simultaneous coexistence of perceptual and theoretical events in the percipient/thinker. Still, the heightening of sensory possibilities offers the possibility of further simultaneous theoretical cognitive activity, as indicated by the following example.

Among the limitations of human visual perception, we are currently limited to a certain more or less shared band of the overall electromagnetic spectrum, typically termed "visible light." We are further limited in that we receive no immediate information as to the characteristics of the light we sense, except what is phenomenally given. There are specialized viewing instruments with various readouts in the field of vision (as a crude example, consider the viewfinder of a camera, which often displays information as to suggested lens opening/shutter speed). Vision that is expanded both as to bandwidth and technical information would promote various sorts of thought as to the phenomena under observation.

Expanding on the above speculation, imagine that humans, or perhaps post-humans, have the ability to magnify images in their visual field, or to abstract color and texture, the better to consider the geometrical aspects of what is presented. Such abilities might prove powerful aids to accompanying thought. Consider the possibility of reversing this process. That is, as one thinks of certain matters visual images can be summoned and positioned as needed. People generally do this to some extent already, but suppose this process of visualizing at will were greatly enhanced as to clarity and detail. Such controlled imaging would likely be the opposite of the hindrance to thought that sense perception has often been said to be.

The difficulties surrounding such speculation should not be ignored. It remains unlikely that various forms of intense thought will be performed in the company of ordinary sense perception. Complex mathematical calculations demand a focus which dismisses perceptual accompaniment. Recall of a theoretical problematic in the midst of a perceptual situation tends to drive the thinking subject to seek a place shielded from the demands of perceptual focus, the better to consider the recalled matter. Yet even these examples might admit speculative reply, particularly if one regards the suggestion of the preceding paragraph regarding "controlled imaging." And to grant that some instances of thought will always be better done by humans in the absence of sense perception is not to disallow the various suggestions that have been presented for the enrichment of perception.

We do not yet possess the deep secrets of cognition in general, and perception in particular. The sciences which include human visual perception in their domain have much to offer now, but many outstanding questions remain. The limitations of future human visual perception cannot be specified. There are well-understood limitations regarding such matters as the dimensionality of the visual field, but even such limitations can be variously overcome to some extent. For example, the limitation of the visual field to two or three dimensions can be transcended by stroboscopic imaging. "Flicker" superposing might allow the visualization of yet more dimensions.

What we may learn about our cognitive systems as they now operate, and about the possibilities of enhancing them, will likely determine a re-evaluation of our various philosophical thoughts, even as it revolutionizes our lives.

NOTES

1. For example Richard Rorty, "The World Well Lost," *Journal of Philosophy* 69, no. 19 (October 1972): 649–65.
2. Descartes, *Meditations*, in *Discourse on Method and Meditations,* trans. Laurence Lafleur (Indianapolis, Ind.: Bobbs-Merrill, *1960).*
3. Henri Bergson, *Introduction to Metaphysics*, trans. T. E. Hulme (New York: Putnam, 1912).
4. Benjamin Lee Whorf, *Language, Thought, and Reality: Selected Writings* (Cambridge, Mass.: MIT Press, 1956).
5. Carlos Castaneda, *A Separate Reality: Further Conversations with Don Juan* (New York: Simon and Schuster,1972), 37.
6. Castaneda, *A Separate Reality*, 23.

Chapter Seven

Alone and Without Love

Human beings are considered social by nature, somewhat in the manner in which lions are considered social by nature, whereas tigers are regarded as not social by nature. As with many of the vertebrates and various invertebrates, humans are sexual in nature, in the sense that they at times desire sexual intercourse with another of their species. Unlike tigers, humans need to be social, in that with regard to social interaction if each human behaved as tigers behave humanity would cease to exist. Evolution has not equipped humans for solitary life. Until recently in human history, sexual intercourse was requisite for human reproduction. On the large scale of species survival, sexual intercourse is still generally requisite for human reproduction, given that comparatively few humans have access to other means of reproduction.

These two characteristics appear necessary features of humans. Yet there are humans who live solitary, isolated adult lives, and there are humans who refrain from sexual activity. If being social and being sexual are necessary for the continuance of the human species, they are not necessary for every human. Which prompts the question: What if at some future time being social and being sexual were not necessary for the survival of the species? Could not the effort humans make in such behavior then be put to better use?

One ready response to this question is that humans do not engage in social and sexual behavior solely because of an evolutionary instinct for survival of the species. They also typically enjoy being social and being sexual. In the form of love of one's mate and of one's children, humans have combined the social and the sexual in a manner largely unduplicated among the other species. Unlike other animals, humans have historically refined these behaviors in numerous ways, influenced by circumstances and their desire for enjoyment. Hence, to ask if humans would fare better removed from society and sex seems to be asking if humans would fare better defying both the evolutionary tendency for species survival and the desire for well-entrenched forms of enjoyment.

A concern with definitions, or essences, prompts the question as to whether beings sans both social and sexual characteristics should be considered human. The odd case of the individual hermit-ascetic is conceptually tolerated by virtue of its oddity. But if anything approaching a species-wide foreswearing of social and sexual behavior were to become the case, the continued application of the term "human" for such creatures would be problematic. In keeping with the aim of this chapter, which will examine the possibilities of such transformation, definitional matters such as this are of little concern. Whether the beings lacking such characteristics are deemed human or not, they originate from humans. The speculative question posed is this: Ought humans, as presently constituted, evolve into such beings?

This question requires refinement. Is the evolution of present-day humans to these other beings an abrupt change? Can the overall question be separated, so that present humans are regarded as changing to nonsocial, sexual beings (like tigers), or to social, nonsexual beings (like no mammalian species on earth)?

Large-scale abrupt change is often accompanied by suffering. In the interest of regarding the resultant beings in their new form, let the distraction of contemplating the suffering brought about by change be minimized, so that the transition is considered to have occurred as calmly as possible, with the evolved beings accustomed to their lives. In a manner, this recommendation is similar to that of not speculating on what it would be like for humans (say, Europeans) to move from their twelfth-century modes of living to late twentieth-century modes of living in the space of one year. Were this move somehow done, one can imagine the sense of dislocation and consequent suffering on the part of these twelfth-century people faced with so abrupt a change. Yet the question of whether (European) humanity is largely happier in their present mode of being than their twelfth-century counterparts is occasionally considered, absent the need to consider whether present people would be happier abruptly forced to live in the twelfth century, or conversely.

There is no reason to prohibit the splitting of the question, so to consider evolved nonsocial, sexual beings, or social, nonsexual beings. Given the psychological and sociological complexities of these questions, it should not be expected that the results of these latter questions simply conjoin to supply an answer to the original question of the possibility and desirability of the transformation of humans to nonsocial, nonsexual beings. With this caveat in mind, these split-alternative questions will be considered prior to the more radical question of the desirability of the full alternative.

Solitary sexual creatures, such as tigers, require large areas without the presence of another of their kind. This widely known fact might influence one's initial thought regarding the possibility of humans leading sexually active, nonsocial lives. Given human population density on earth, one might suppose little possibility of widespread, nonsocial nature for "evolved" humans. Of course, population density could be significantly lowered by various drastic means. There is the generic, post-nuclear-holocaust scenario in which mutant humans scavenge amidst the ruins in greatly reduced numbers. The desirability of such lives is in question. Other means for reducing the population in a controlled and gradual manner can be contemplated. Such means, together with results, are not as prima facie undesirable as post-apocalyptic versions of reduced populations.

Alternatively, and contrary to initial intuitions, one might contemplate a world in which humans have evolved into nonsocial beings living in population densities similar to, or even greater than, what is currently the case. Since this supposition appears factually impossible, reasons for entertaining it should be given. Two such reasons come readily to mind. First, on the supposition that such beings are sexually active, they may reproduce at rates which will keep population levels high. Second, maintaining desirable living standards might require a large population, engaged in aspects of production and distribution of goods and services.

The latter reason immediately points to the seeming factual impossibility of the assumption of evolved nonsocial humans living in circumstances of current or increased population density. At such population levels, humans cannot live solitary hunter-gatherer lives, and hence must depend on the cooperation of others for the successful appropriation of material reality necessary for their mutual survival. Such cooperative dependency is social by definition.

The above reasoning apparently puts an end to all such speculation save that regarding the case of the post-human at gradually diminished population levels. Such a vision of the future can be made desirable only if the lives of such comparatively widely spaced beings can be shown to be other than that of devolved, animalistic mutations. There are two general versions of such a possibility, which are termed the "high-tech" version and the "low-tech" version.

The low-tech version of gradual depopulation involves decreasing reliance on advanced technology and increased reliance on so-called "alternative technologies." While this alternative seems the most consistent possibility to conjoin with the contemplated future of nonsocial beings, there are problems to be addressed. There are those who currently favor a dismantling, or at least a de-emphasizing, of our current technologically-intensive mode of life. While this longing for simpler, more in-touch-with-nature living has an attractive pull, those favoring it often imagine themselves living such lives as the highly educated people they typically are. The apparent inconsistency is that such education, if it is to be widespread, demands a certain level of infrastructure that likely cannot be maintained in a low-tech environment. Further, while a low-tech style of living may not necessitate large population centers, it may necessitate reliance on smaller social units such as family, clan, or tribe, given the diminished capacities of individuals unaided by advanced tools. While this last observation is not a criticism of the commonly encountered low-tech vision of humanity's future, it runs contrary to the current assumption that the beings inhabiting this future are nonsocial.

Whatever responses can be made to the above critical observations, the low-tech version of the evolved, nonsocial life will not be further explored. The rationale for the desirability of the nonsocial life, to be given below, conflicts with this alternative. Instead, focus will be on the high tech version of this life. The assumption supporting this choice is that technology, appropriately developed and applied, enhances life's possibilities.

The high-tech version of the reduced-population nonsocial life confronts the challenge already given; how can advanced technology exist in a nonsocial environment? The assumptions behind this challenge include child rearing necessitates others, education necessitates others, and research/production/distribution necessitates others. The force of this challenge is undeniable, especially if humanity is considered in its present circumstances. Still, the child rearing part of the challenge admits an immediate response. As with nonsocial mammals such as tigers, the proposition that if reared in childhood by others the adult creature will associate regularly with others is not necessarily correct. There are degrees of nonsociality, and while it is at least logically possible to contemplate an intelligent creature that is nonsocial from birth, there are wide possibilities in the contemplation of the nonsocial maturity of a creature raised by others. Similarly, occasional passing contact with others of the same species is not to be understood here as a social life. Beyond such meager characterization, the notions of social and nonsocial will be left intuitive.

The response to the challenge arising from the consideration of child rearing does not address the large remainder of the challenge against the possibility of high-tech, nonsocial life. How can creatures develop and employ advanced technologies without socially interacting? The very existence of artificial or natural language, generally believed necessary for complex thought, arguably demands communication with others.

The question of the possibility of private language, and the question of the possibility of detailed "sublinguistic" thought, will not be explored here, as they have been widely discussed and are still largely unsettled. Pursuant to the overall question before us, it is assumed that evolved humans either acquire language in childhood, or successfully develop private languages, or employ a version of thought which in some manner is supported by sublinguistic neurophysiological activity.

Large questions remain. How do such evolved humans manage high tech living in a nonsocial manner? How do they pool their knowledge if they are living nonsocially? And, if these two questions are answerable, what is the advantage to living nonsocially?

Present day humans, as well as humanity in the foreseeable future, cannot manage high-tech living (as opposed to solitary, frontier-style living of a bygone era) in a nonsocial manner. Interpersonal communication is necessary in myriad circumstances. Still, humans might evolve into beings capable of absorbing and using quantities of information far beyond present capabilities. These speculated beings might be able to conceptualize, hypothesize, and theorize far beyond present capabilities. Further, their technology might be engineered to support their style of living. This technology might be auto-evolving, and keyed to the wants and needs of individual users. It might have been designed to behave in this manner while humans were still in communication with one another, yet at the threshold of a nonsocial mode of life. The technology might manage such pooling of information as is necessary for individual pursuits. The technology might guard against severe conflicts among individual pursuits.

Obviously, a blank check has been issued to "the technology" to solve major problems. What this technology would be is the stuff of imaginative speculation, which will not be indulged in a manner down to particulars. No one can currently offer details, and no one can currently issue a reasonable denial of the possibility of such technology nor a reasonable denial that a significant amount of this technology would be physiologically internalized by the speculated, evolved humans.

Evolved, possibly post-human, beings have been considered in a depopulated, high-tech environment. Having burdened the high-tech hypothesis with the assumption that concomitant technology solves problems in that environment, there is little to prohibit the following additional assumption. Consider that beings evolved from present humans live in a highly populated environment. Without specifying population density, let it be considered that these beings exist in a population density roughly akin to that in present-day major metropolises, yet live nonsocially. The *prima facie* absurdity of this assumption is, of course, that beings in such proximity could exist without social interaction. Further, allowing this absurdity for the moment, what advantage would such a scheme have over the high-tech depopulated scheme considered above?

To address the latter question first, the quick answer is this: strength in numbers. Even if the beings in this populous environment are not in communication with each other, they might benefit each other in manners having to do with their interaction with technology, which would be mediating their noninteractive lives. In this scheme, the technology through which individuals managed their lives would, in effect, be the overall guardian of welfare. More individuals actively engaged with such technology might lead to enhanced progress along mutually beneficial paths.

This all is admittedly alien-seeming, strange to a high degree. The vision is that these beings live in close proximity to each other, yet avoid social interaction. They do not communicate with each other by means of their technology (for example, by e-mail), yet somehow they are engaged in cooperative living, although the ends of such living have not yet been indicated, and the means of such living have been broadly ascribed to technology. Perhaps the following will mitigate the apparent strangeness. In this vision, people largely do not communicate with one another because such would be a waste of time, even as present day humans typically do not engage in idle conversation while deeply absorbed in tasks. Generally, humans engage in conversation for such reasons as to secure their immediate goals, or for the pleasure of each other's company. Suppose future beings do not need to converse with others to secure their immediate goals, and such enjoyment as they receive is not gotten from the company of others.

It should be noted that something akin to enjoyment, perhaps directly akin, is being postulated for these assumed beings. If they take no satisfaction in their lives, if they operate in the manner of unfeeling beings who are nonetheless flexible, adaptive, and goal oriented, then it is unlikely that any case can be made for humans to evolve into such beings. To note this is to raise broad questions of ethics. Such questions would be pursued in detail, if one believed a satisfactory resolution within the tradition were possible. Is hedonistic utilitarianism being proposed here? There is an element of such, but to reverse the question, assume a society of beings that successfully behave as Kant wants his ideal society of autonomous willing agents, respectful of the moral law, to behave. Make the further assumption that to secure this society such agents must not take pleasure in the acts they perform. This is not an assumption Kant will grant, yet Kant's morally ideal society likely cannot be realized among presently constituted humans who seek their "heteronomic" enjoyments. Conflicts will inevitably arise. Kant is doubtless aware of this, and would reply that it is the responsibility of the individual agent to place duty above enjoyment in cases of conflict. Such enjoyment as the moral agent receives from being moral is independent of the agent's morality. This well-known scheme strikes some as so alien, so inhuman, as to be otherworldly.[1] To the extent that one is so struck, one will allow for the enjoyment of one's activity as integral to a good life. Which allowance is insufficient by itself either to deem one a hedonistic utilitarian, or to entail that one finds little of value in Kant's moral theory.

Prior to facing the question of what such nonsocial, evolved (post?) humans will be pursuing, or what about their lives is enjoyable, the question of their sexuality remains. In positing their nonsociality, it has been assumed that these beings do not form families, and refrain from various forms of loving contact with one another. Let it be further assumed that these beings lack sexual urges. In response to the obvious inquiry as to why present-day humans might desire to evolve into such beings, it will be argued that sexual desire is both unnecessary and undesirable in such beings, who nonetheless lead desirable lives.

Sexual desire in such beings is unnecessary to the goal of procreation, which from an evolutionary standpoint is the explanation for present-day humans having sexual desire. Two reasons support this judgment of the nonnecessity of sexual desire for our contemplated beings. First, if procreation is necessary for the continuance of the species, this can be accomplished in other ways. Second, as has been discussed previously in "Facing Immortality," these evolved beings might lead indefinitely extended lives. Accordingly, such beings might not desire progeny.

It is not possible to state categorically that sexual desire is unnecessary for such evolved beings. It has been argued that such desire is not necessary for reproductive purposes, but other necessities might be present. If such necessities are the case, then such beings will be sociable to the extent that their sexual desires need be discharged through congress with another. This grants such beings a minimal amount of sociality, on the assumption that courtship rituals are minimal. Further, if sexual desires can be satisfied by individuals acting individually, then such necessity as might exist does not entail social contact.

Sexual activity is a source of pleasure for present-day humans, and for much of their lives sexual desire is pervasive. Why subvert the desire, and for what reason forego the pleasure of its satisfaction? Powerful reasons have been given, by Plato and Freud for example, for keeping sexual desire as an underlying source of energy for individual and social transformation. Yet it is undeniable that people are occasionally misled by their desires. As Plato notes in the *Protagoras* (356–358), people may mistake a present pleasure which brings pain in its train for an absent, but longer-term pleasure, burdened with less pain. For Plato, the immediate presence of an intense desire is not sufficient reason for acting on that desire. Still, at least in the *Protagoras*, Plato apparently accepts the hedonistic calculus as an appropriate action guide. In other Platonic dialogues, Plato is unwilling to accept pleasure as the appropriate goal of human activity.

If humans evolve into something other than they are now, it cannot be presumed that desires, even deeply rooted ones, will accompany them in their transformation. If procreation is not a reason for these speculated future beings to retain sexual desire, then it is either the biological fact that such desire is deep seated within us, or it is the pleasure that fulfillment of such desire brings, that constitutes a reason, or reasons, for its presence. It is possible that in the passing of time our biology will change to the extent that we no longer have sexual desire, nor do we derive pleasure from sexual activity. On the assumption that humans gain sufficient control over their future evolutionary possibilities, the question arises as to the desirability of transformation into beings lacking sexual urges.

We should remind ourselves that there are differing views as to the importance or desirability of our sexual urges. If one is stationed in a middle position here, as are likely the overwhelming majority of humanity, one's immediate thought will be that sexual desire and activity are normal, pervasive, and perhaps fundamental parts of being human. There are those, however, who do not share this belief. Plato obviously struggles with the issue of evaluation of the place of sexual desire and activity in human life. The Catholic Church can be considered as having a tensioned position on the question. There are examples of asceticism from a variety of cultures, such examples often highly regarded by those of the same cultural background.

One might devalue sexuality for religious reasons, thinking sexuality to be in some manner a profanation of the sacred. One might think that sexuality is in some other-than-hygienic manner unclean. Then there are well-grounded hygienic considerations leading people to devalue sexual activity, and the desire from which it arises. Or, as with Plato in some of his writing (particularly the *Phaedo*), one might devalue sexuality because it conflicts with one's desire for knowledge. Indeed, those parts of the Platonic corpus in which sexuality is not under attack, it is either enlisted as an aid in securing knowledge (as in the *Symposium*), or accepted as a fact of human nature, to be properly positioned so as not to interfere with the ultimate quest of humans, which is knowledge.

Humans have struggled against nature, including infirmities in their own nature, and have struggled against each other. It may come to pass that both these sorts of struggles are minimized. Whatever humans evolve into, including the possibility that biologically they remain as before, they may achieve a mastery over various external and internal natural calamities. They may learn to live non-combatively with one another, this latter possibility having some relation to the possibility of mastery over various calamities. Of course, neither of these possibilities seems realistic at this time. External nature overwhelms us in ways we can at best mitigate, but can neither prevent nor forestall. And as long as we interact with one another, we will seemingly have conflicting needs and desires.

While some events in nature will almost certainly be threatening regardless of whatever humans evolve into, it may be possible in some future time virtually to assure continuance of an individual through measures which either are unknown now, or cannot be implemented in the foreseeable future (some such speculative measures are mentioned in "Facing Immortality"). Regarding our conflicting needs and desires, there are two general strategies for alleviation. We might evolve into beings that interact with each other in a thoroughly cooperative manner. Alternatively, we might evolve into beings having little need for contact with each other.

Both these speculative alternatives appear fantastical. Yet the idea of human or human-like beings living cooperatively perennially reappears in religious, political, and philosophical thought. The pessimistic observation that this ideal remains unrealized, and that attempts at its realization have at times been catastrophic, is not merely historically accurate. It is also correct to note that no detailed plan for securing such cooperative existence (in the large among nations and societies, or in the small among handfuls of individuals), has secured widespread acceptance. Given present knowledge and resources, it may be possible to form small groups of completely cooperative humans, but at the cost of both questionable means and ends. For example, a group of people kept extremely tranquilized, with overseers monitoring and attending to their needs as required, would plausibly interact cooperatively, or at least noncombatively. Beyond such distasteful possibilities, any optimistic scheme appears to entail alterations in human psychology going beyond current regimens of tranquilization or behavior modification. There have been periodic discoveries of people living in primitive, tribal manners, gently and with near-complete cooperation. Some of these ostensible discoveries have been debunked. In other cases, as with some Eskimo societies not yet influenced by outsiders, such cooperative living has been verified. Yet few people in developed society wish to live in the circumstances of the nineteenth-century Eskimo.

In speculating on the possibility that humans evolve into nonsocial beings, albeit in significant population density, such seeming anomaly eased by technological possibilities, it must be allowed possible that humans evolve into social beings whose fully cooperative lives are secured at least partially by means of technological developments. The overriding question is why preference a nonsocial scheme to a social one? While events may transpire such that one or the other of these schemes is realized in a currently unpredictable fashion, for now there is room to speculate. If speculation reveals one of these schemes preferable to the other, then humans may be able to orient themselves as a species towards the implementation of the favored scheme. This last observation invites the question: preferable to whom? To present day humans? To sometime future still-humans? Or to sometime future beings in the process of evolving, in a planned or unplanned manner?

These are not easy questions. For most people, likely including many philosophers, these appear to be avoidable questions, in that there is no pressing need to address them, and the context in which they are posed is wildly speculative. There are those, however, who in one manner or another believe humanity is in, or near, a state of crisis. These include those concerned about population growth, pollution, global warming, famine, epidemic, resource scarcity, weapons of mass destruction. They may be mistaken on all these points. Granting the possible realization of one or more of these problems, humanity has survived various such catastrophes—although not without attendant large-scale suffering. One could argue that given our numbers, and our reliance on increasingly sophisticated and interlinked technology, any such global misfortune would have greater species-threatening consequences than in the past. Alternatively, one could argue that present human know-how, combined with the number of enabled human survivors, would ensure a revitalization of human civilization at a technologically advanced level.

Most people do not have a sense of urgency regarding "doomsday" scenarios. If any such circumstance were a near-term likelihood and perceived as such, it might be too late for humanity to take action, even if a course of action were largely agreed. It is in any case unlikely that the majority of humanity would agree to some sort of controlled evolution unless a perceived general threat of some sort were believed to necessitate a radical response. The sort of change in human nature under examination will not be sold to humanity on the basis of its promise of a better life than the generally perceived less-than-threatened life many people currently possess.

It thus appears that the sort of evolution of the human species that has been contemplated above will not occur. The reasoning to this conclusion is probabilistic, and the evidence upon which the hypothesis of such evolution is measured lacks sufficient weight to assure us that we or our descendants will not confront the decision. Given the roughly one hundred millennia in which the human species has existed in its current biological form, it seems a safe assumption that it will be as biological humans that the choice will be initiated to evolve in a controlled manner, if such a choice is made. We may make the choice only if scientists show that the choice is reversible, should it be generally perceived a bad choice after implementation. We may take slow steps, first expanding our cognitive capabilities in various manners without undue alteration of our overall physiognomy. If our evolved descendants continue making evolutionary choices, what we may become one hundred millennia hence is subject to sober discussion in only the most general terms.

Speculation about what we may become at times leads to nonbiological replacement of aspects of our physiology. Many have an intuitive revulsion to the proposition that we transform ourselves to robots. Yet we willingly replace various parts of ourselves with artificial devices (our teeth, for example). Nonbiological replacement may reach a point at which we no longer regard ourselves as primarily biological entities. Alternatively, we may alter our biology significantly, while remaining largely biological beings.

Arguably of more interest than general speculation on the physiological transformations humans may undergo in the following millennia is speculation on transformations of our psychology. It seems likely that we will discover increasingly powerful means of improving various cognitive abilities. The sociological ramifications of these discoveries include the areas of discussion in this study. If we become orders-of-magnitude more cognitively potent than we are currently, there is little doubt that our current modes of living will be significantly altered. The pre-Socratic philosopher Xenophanes was noted for saying that the various animals, if they could, would make images of deity in their species' likeness (e.g., horses would draw the gods as horses).[2] One might pursue this thought by wondering how, say, dogs would behave if suddenly endowed with human cognitive abilities while in their biological dog physiognomies. A whimsical speculation to be sure, but if one rides with it momentarily it leads to the point to be made regarding our own evolutionary future. "Superdogs" would be hampered by their inability to grasp things (that is, physically grasp things), their initial lack of vocabulary, and their vocalizing limitations. Empowered as speculated, however, they might move to overcome these limitations. If they moved in this general direction, they might seek physiological changes by some means, and they would likely modify their doggy behavior to a large extent. Lacking as this quick speculation is, it underlines the point: why should current humans suppose that their distant descendants will fundamentally resemble them in the manners in which, say, seventeenth-century humans fundamentally resemble us?

Predictions and speculations regarding future specifics are notoriously inaccurate. Attempting to see 100,000 years ahead, even in generalities, invites present-time ridicule. Yet the questions, once contemplated, are intriguing. What will our beyond-the-foreseeable-future descendants be like? How will they live? Will they continue to have "primitive aspects" such as potentially harmful emotional responses, unregulated appetites, irrational fears? Will they continue to delight in various activities? Will they be largely supportive of one another? Will they form small social units, larger (less than species-encompassing) political units?

These and other such questions are obviously unanswerable with certainty, or even respectable probability. Such uncertainty being the case, one might question the utility of over-the-horizon speculation of post-human, nonsexual and nonsocial existence. Such questioning is typically founded on two concerns: one, the value of utopian musings given current pressing concerns; two, the low probability that any such presented scheme will correspond to actual future states of affairs.

Utopian and futurist thinkers have often bemused sober thinkers, and at times have prompted sterner responses than mere bewilderment. While occasionally celebrated for their forecasting, the more famous among them, such as Jules Verne, have usually gotten the details embarrassingly wrong, and in their inability to foresee have omitted revolutionary developments that pervade contemporary life (for example, computers, atomic energy, jet-turbine engines). One moral of this is that speculation on the distant future should be limited to broad generalizations. But why such speculation at all? Because as a species we have arguably only begun to realize our potential. We have realized much in our brief moment of civilized life. Yet we are beset with problems, both human caused and external to us, that diminish the lives of many of us and sometimes threaten the survival of the species. To an extent, these problems are addressable by current means. However, we are not yet in position to negotiate many of these difficulties because we lack both understanding and material means for coping with them. Regarding such difficulties, it is perhaps not out of proper order to offer idealized speculation as to future possibilities.

Our distant descendants might continue to enjoy the company of others. Their lives might continue to be of strictly limited duration. Their lives might continue to be staged as infancy, childhood, adolescence, maturity, and senescence. The greater part of their lives might be strongly influenced by their sexuality, as with our lives. Although linked, these circumstances can be varied to some degree. Perhaps sexuality disappears. Or perhaps there are fewer stages, or considerably longer duration of life.

One possibility from this palette: our distant descendants do not enjoy the company of others, they live indefinitely extended lives with fewer stages, and are not sexual beings. This possibility is given some support by the following considerations. If the human species and its descendants should continue for, say, one hundred millennia, it is likely that they will have learned to live in a more peaceful and mutually supportive manner. Alternative possibilities must be acknowledged, but there is the chance that our distant descendants will have solved the problem of peaceful coexistence. Again, although alternative possibilities exist, let it be supposed that these descendants are scientifically and technologically at a great advance of the present state. These suppositions granted, it is possible that the contemplated descendants will take enjoyment in discovery, that their self-satisfaction will be as learners and knowers. It is also possible that they will not need the company of others in pursuit of such satisfaction.

This is not to say that such speculated beings will not welcome, and perhaps need, the contributions of others of their kind. They might work cooperatively on various ventures. What is being considered here is that they might not desire to have social or sexual contact with each other. Their contributions might be mediated, for example, by devices descending from present-day computational machinery. Similarly, they might communicate with one another in some such manner, not conversationally, but in an impersonal style of sharing or pooling information.

None of this is to suggest that these beings will not care for each other. That they do not seek the company of others does not entail that they do not value one another's well-being. Perhaps they do so strictly for self-interest, aware that others are of help to them in their ventures. Perhaps, however, they desire the well-being of others because they delight in one another's good fortune. Such delight is not inconsistent with the nonsocial, nonsexual existence ascribed to them. It is possible that along with solving many of the problems of scarcity, distribution, and general welfare that currently beset humanity, these beings have evolved to the point of leading lives that a moralist such as Kant would regard highly.

Some believe that competition fosters gain, in that a society of satisfied, well-provided individuals is a society whose members are not impelled to produce in a manner that might be of benefit in unforeseen future circumstances. If the contemplated descendants of present humans are well provided, and not in competition with one another, then such a society (of assumed nonsocial members) is arguably stagnant. Of course, one person's stagnation is another's lack of turmoil and dislocation. Waiving the last remark, it is not clear that the contemplated future society—sketched in the briefest terms—would fail to progress. One sort of competition is interpersonal, and athletics often serves as a model for such struggle. Yet even within athletics there are instances of competition that are not directly interpersonal. Consider a rock climber who sets the task of climbing a rock wall. This could be in the context of competition among a group of climbers. Alternatively, the contemplated climber might be measuring himself or herself against the task of climbing the wall, with little thought of anyone else who might someday attempt the task. If the contemplated climber has not previously climbed the contemplated wall; if, say, the climber is not confident that the task is within ability; if the climber feels that the attempt will likely improve present skills, then this is not a case of resting at one's present station.

The analogy of the solitary climber with the contemplated solitary future member of nonsocial society is patent. The evolved being of the future seeks increased understanding. This is not gotten by reviewing knowledge already gained. The contemplated future individual is confronted with the multidimensional rock wall of ignorance, and must assess ability and means of climbing this wall, after which other walls are to be surveyed.

It is difficult to imagine evolved humanity (or post-humanity) as a collection of solitary seekers of knowledge. Despite suggestions given above regarding impersonal communication of research, we presently regard knowing as an essentially social enterprise, involving the social aspects of child rearing, education, research teams, and so forth. Perhaps the following example will mitigate some of this difficulty. Suppose a future being researching high-energy physics (assuming there remains this general field of study) needs to run an experiment on a circular collider having a diameter of one hundred kilometers. Imagine further that no such apparatus is available. In present circumstances, that would be the end of the matter, unless the scientist persuades the powers that be to commit resources to the construction of the device. Given the scale of the device (somewhat larger than the recently proposed superconducting supercollider), the project would involve many people in the implementation and operation of the device. That's now. Perhaps 50,000 years from now, if there is need of such a device, an individual requests it through a centralized system, whereby machinery constructs the device in a few hours, runs the tests and dismantles the device, restoring the environment of the device to its previous state. How this would be done is beyond present understanding. That in 50,000 years it could not be done in the general manner described is a proposition to which probabilities cannot legitimately be assigned at present. Allowing one's imagination to be limited only by presently understood universal constants of nature allows much leeway in such speculation. Granted, if these descendants of present humans share much of their ancestors' psychological constitutions they will want to extend the frontiers of their knowledge, which are frontiers by virtue of their relative inaccessibility. These extensions may require a longer-term mobilization of resources. The potential for competition is apparent, together with concomitant politicking for one's individual project. Such politicking is by definition a social matter. That is how things stand now. Speculation liberates one from the necessity of the assumption of human nature as it is currently manifested. If humanity, or post-humanity, survives and flourishes for 50,000 years from now, our distant descendants may well have conquered envy and competitive urges to the extent that the need for socialized decision making no longer exists.

If the lives of our distant descendants are governed by some concordance such that they are mutually supportive, rather than competitive, yet not needing or desiring daily social interaction, and if these beings are driven by desire to know, sharing their knowledge "conquests" through a form of information pooling, then the distant future of humanity will differ radically from its past and present. These individuals may have emotional natures markedly different than ours. Much of our evolutionary inheritance might be altered or discarded as being not needed, or undesirable, under then-prevailing circumstances.

As presently-constituted humans gain increasing control over their own physiological possibilities, including such cognitive and conative aspects as are supported by their physiology, they may seek a direction of evolution indicated by their aspirations. Weighing the balance of sensual joys against the pains of frustration, envy, and competition, they may seek to modify themselves so as to improve this balance. Alternatively, they may reject the entire matrix of sensual pain and pleasure, for reasons of conflict with goals held in higher esteem. Along with this rejection, they may not have need or desire for one another's company.

NOTES

1. Nietzsche in *Twilight of the Idols*:

> Progress of the idea: it becomes more subtle, insidious, incomprehensible . . . The idea has become elusive, pale, Nordic, Königsbergian.

The Portable Nietzsche, trans. and ed. Walter Kaufmann (New York: Viking, 1954), 485. The "idea" here is that of the "true world." I do not doubt that Nietzsche would apply these same words to the evolution of the concept of morality from its beginnings to Kant. As he says in *The Antichrist*, section 11,

> "Virtue," "duty," the "good in itself," the good which is impersonal and universally valid—chimeras and expressions of decline, of the final exhaustion of life, of the Chinese phase of Königsberg.

Nietzsche, *Portable Nietzsche*, 577.

2. See Xenophanes, fragment 15, in Kathleen Freeman, *Ancilla to the Pre-Socratic Philosophers* (Cambridge, Mass.: Harvard University Press, 1956), 22.

Chapter Eight

Temporarily in Conclusion

I have largely avoided the personal pronoun in much of this book. I would continue to avoid it, were it not that I feel a pressing need to make an autobiographical statement. I fear being misunderstood, taken in an unnecessarily bad light, should I have the fortune to see this work published.

The history of philosophy contains various cases of philosophers publishing thoughts which inflamed the intellectual community. Spinoza sought anonymity in publishing his *Politico-Theological Treatise*, although such was denied him. Hume exhibited his characteristic wisdom in arranging for the posthumous publication of his *Dialogues*. It would transcend arrogance for me to place this work in the company of those giants, but I worry that whatever distribution this work has, there will be proportionate misreadings and speculation on the misanthropic nature of the author.

There is a sometime principle to which philosophers are trained to adhere: judge the thoughts and not the character of the author. Were this principle universally followed, my autobiography would be unnecessary. One need only reflect on Nietzsche's celebrated attacks on Kantian ethics via the character of Kant ("the Chinaman of Konigsberg"!) to remind oneself that focus on ideas to the exclusion of focus on the writer is an occasionally-violated principle. I would like to deflect such wayward focus.

So who am I?

I have been a philosophy professor for about thirty-five years. I was trained at an analytic graduate program. My specialties are metaphysics (my dissertation was on the subject of universals), epistemology, and philosophy of mind (focusing on philosophical issues of the artificial intelligence debate). I have published in these areas, and have occasionally published in the areas of ethics and social philosophy. I teach at a mid-size, Midwestern university.

I have been married for many years, and have two grown children and three grown step-children. I have an intense love for my family. I have pets and tend to treat them as members of the family. I have only a handful of close friends, but I treasure them. There are others whom I regard in a friendly manner, but with whom I would not share personal details of my life.

I am not inclined to asceticism, isolation, misogyny, misanthropy, or dislike of children. I have optimism regarding humanity's near-term prospects. I may not be the person you would suppose to be the author of this book.

Why philosophize like this?

I do not claim that speculative philosophy of the foregoing sort should form the bulk of philosophical effort, and I would not argue that its significance demands a sizable minority of practitioners. Philosophers tend to fasten on interests with an intensity approximating that of practicing scientific researchers. A glance at the relevant trade publications, such as the *Proceedings and Addresses of the American Philosophical Association*, especially in the listed programs for any of the three divisions, indicates current interests. A retrospective examination of such programs indicates the duration of those interests. One might inquire as to the relevance of much of the current discussion in English-speaking philosophy as easily as such inquiry is made regarding my thoughts.

The question remains: Why should anyone philosophize in this manner? Is a new direction for philosophy being suggested? Yes, a new direction is being indicated, but not at the expense of prevailing interests. This is simply to recognize the fact that those committed to a research program, with attendant professional success, will likely adhere to that program until their retirement. Kuhn's observation that complete paradigm change comes only with the dying of current practitioners applies here.[1]

I believe there is a historical comparison suggesting that the sort of speculative philosophizing I have been doing is not novel. There is of course the general tradition of utopian literature, but in various of its exemplary works this tradition has had a theological cast. I see my work as more in the tradition of Plato's *Republic*, cognizant that Plato's idealized city-state is itself regarded by many as a utopian vision. So regarded, I am partially transferring the question as to why I am philosophizing in this manner to the question as to what Plato had in mind in writing the *Republic*.

It is appropriate to venture that Plato had several things in mind in writing the *Republic*. I identify one of his ostensible aims, the betterment of society through appropriately applied knowledge, as emblematic of my effort. Not that I think society can be perfected. But neither did Plato, who has Socrates argue (472a–473b) that an approximation to the ideal is the best for which one can hope. Unlike Plato, I do not suppose an ideal endpoint, but rather an evolution towards a better state of affairs than is current. "Better" by the judgment of those who look back on our current existence, as we look on the existence of those in some earlier times and pronounce our lives to be on the whole better. There is much that can be said here (and this topic was discussed in "Facing Immortality"). Not everyone thinks the life of a feudal serf, or knight, or mistress of the castle, was of lesser enjoyment and self-satisfaction than present lives. Still, there is the impulse of humanity to absorb new knowledge and implement new technology. Not everyone shares this impulse today, and qualification is in order, especially regarding the implementation of new technology. Plato's idealized society itself seems static in this regard.

A further unlikeness: Plato gives the impression that the Republic is within reach, if only fortune shines on human affairs to the extent that the proper steps are taken. He may not have thought this, but it is easy to read him in this manner. I think it will take some time to evolve into what may be a posthuman future. Yet with these differences, I still regard my work here as in the Platonic spirit of looking to the better, if not the Good.

Critics of the *Republic*, and of proposed utopias in general, argue that such thinking does little to alleviate current social maladies. Telling the Athenians that life will be better for all once the philosopher-king(s), who know the Good, are empowered, invites the question: given that no one seems to know the Good yet, what is to be done in the meanwhile? Plato is in a better position to respond to this question than I am, if he is assuming that the knowledge sought is near at hand. I do not foresee a knowing of the Good, and I am occasionally speculating in terms of hundreds and thousands, perhaps tens of thousands, of years.

Humanity might confront some of my speculative possibilities in the nearer term. It might not be long before scientific research yields the option of indefinitely prolonged lifespan. In the midst of worldwide poverty alongside comparative wealth, how would this option be parceled? Is it premature to devote philosophical thought to this question of distribution? Are questions of distribution sufficiently generalizable that overall theories of distributive justice encompass the more specific question of the distribution of longevity, or are there some distinctive aspects to this question? Distribution schemes typically assume normal human lifespan, so that any scheme for the distribution of longevity might need consider the ramifications of such distribution within a population heretofore strictly limited in this respect.

Here is another consideration regarding my speculations. If you are reading this, you are most likely well educated and not materially impoverished, relative to various other humans. I ask you to perform an imaginative feat. Imagine that the overwhelming majority of the world's population are living at similar intellectual and material standards as yourself. If you travel to the interior of various continents, if you sojourn in what are now the zones of impoverishment of various metropolises, if you visit the small islands of the Caribbean, you encounter only people as educated and materially well off as yourself. Cultural differences remain, except insofar as they reflect and support ignorance and poverty, such conditions (i.e., ignorance and poverty) being determined by standards you would apply to citizens amongst you. This last is not cultural myopia born of arrogance. It is possible to preserve some cultural identities and differences while standardizing health, wealth, and education. My own Jewish tradition is exemplary in this regard. While fears of homogenization are not unfounded, any educated and cultured reader should consider the varied cuisines, art forms, and general experiences had amongst similarly educated and prosperous residents of foreign cultures.

Having imagined such a world situation, consider the following questions: On the whole, would such an imagined state-of-affairs be better than the current one? How would this change be brought about? And, most important with respect to my intentions here, where would we go from there?

I think the first question is the easiest to answer, but it is not an easy question. My temptation, and hopefully yours, is to respond that such a world would be a better world than the one we currently inhabit, because social inequality would largely be banished. Social inequality is best understood as unequal distribution of material wealth and social services, rather than in terms of access to venues such as private clubs. Many people long for such equality, but often this longing is expressed as a desire to see a leavening of wealth, so that inhabitants of the developed sectors are urged to lessen their consumption and general living standards, with consequent resource savings being distributed among the underdeveloped. I shall not argue what I think about this, that it is a misdirected scheme for achieving social equality. Instead, I merely note that I think a better result would be the raising of others to my standard of living.

That is my easy answer. It invites familiar criticism. Much of my lifestyle is held to be unnecessarily resource depleting. The world population could not generally be raised to my standard of living without exceeding limits of environmental degradation, so as to summon ecological disaster. Many people among the so-called "underdeveloped" do not wish for my lifestyle, even in the most general terms in which it might be represented. These, and other well-taken criticisms, make the first question a difficult one. I do not think such objections are without telling responses, but I shall not attempt them here—for adequate replies to these objections, if such replies are possible, would demand an entire volume.

As to the question of means, how such change will be effected, I propose to say nothing. I am ignorant as to how this change, which I regard as maximally desirable, can be made to happen. Far from thinking this change impossible I think it quite likely, as humanity becomes more self-conscious of its history and potential.

Let us now suppose that the change has happened, as it well may in the near term. What then? I believe that as a species, rather than as a nationality, race, or creed, we would focus on expanding our possibilities. At this point, relatively unhampered by sociopolitical-economic crises, we would devote some amount of our comparatively enhanced material and intellectual resources towards the sorts of areas I have been discussing. In what manner "towards," and which "areas"? I expect that we would consider those areas that seem most easily attained and more worthy of attainment. Extended longevity is a more likely prospect, at least at this point, than parallel consciousness. Whether such longevity is of more worth than such consciousness is not as easily decided. We may decide that we do not want indeterminately extended lives, even if such a possibility is within implementation. I have argued in "Facing Immortality" that we will likely choose such lives, but I am aware that this conclusion is highly speculative. In any case, I believe there will be decisions about the allocation of resources for at least some of the above proposed transformations of us.

We may of course be confronted with various possibilities absent the sort of equality proposed. In either case, I am unprepared to speculate as to the manner of "moving towards" implementation of any of these possibilities. It would be nice if everyone could be offered them, that they could be effected with a minimum of suffering, that they prove worthwhile in that our descendants could in some neutral or near-neutral manner regard their circumstances before and after and prefer after; and that if their preference is before, means are available for return to before with a minimum of suffering occasioned by the transition.

I am interested in the sort of equality I proposed above for the reasons that it is morally indicated and that it is a likely possibility as humanity emerges from its primal struggles. But isn't this sort of equality the "heaven on earth" alluded to in "Revolutionary Ethics"? The ostensible problem is that people would not desire to change circumstances if they were living in heaven-on-earth. Yet I have indicated that people having achieved the high level of equality sketched above would then seek changes in their circumstances. While I did not intend that such people would seek to unequalize, either directly or indirectly, I did intend that humanity, equalized at a high level of goods and services, would seek various profound alterations.

The heaven-on-earth referenced in "Revolutionary Ethics" is not achieved simply by equalizing worldwide living standards at a high level, rather than at diminished consumption. A substantial part of that scheme includes theoretical knowledge of human behavior that is as yet unavailable, and would not necessarily be available even if living standards were equalized as proposed. In contemplating humanity's desire to alter circumstances in certain radical manners, we have thus to consider the current human condition, the condition of "high level" equality minus other factors designated in "Revolutionary Ethics" as realizing heaven-on-earth, as contrasted with the condition of humanity with those other factors. There are obviously other possibilities. Plato's proposed ideal state seems his idea of heaven-on-earth, yet it is founded on supposedly innate differences among its citizens necessitating, according to Plato, radically unequal modes of living. I ignore this possibility for the reason that I do not accept Plato's premise regarding the necessity of innate differences in mental capacity. I dismiss other possibilities because I think humanity is vectored towards the equality I propose. I grant again the speculative nature of this assumption, but for purpose of discussion I will pursue the consequences of the assumption.

At present, there are many pressing demands on humanity. Scientific discoveries leading to enhancements such as increased intellectual capabilities, or enhanced longevity, might prove as much a burden as a boon. As an instance of burden, the overall consequence of enhanced longevity, were it available now, might occasion further envy and resentment, given the likely limited distribution of this enhancement. The current general condition of humanity argues for resource allocation directed at alleviation or mitigation of current ills, rather than directed at research into possibilities of enhanced longevity, parallel consciousness, and so forth. Nevertheless, discoveries often come as surprises to all but the relatively few directly engaged in the research from which the discoveries emerge, and some of the speculations here may be realized before we are adequately prepared for them.

It is my belief that humanity will someday largely acknowledge the desirability of the sort of equality I have sketched. Some may claim this acknowledgment is already the case. Even if this is so, it seems undeniable that such equality is not sufficiently prioritized as to promise its achievement in the near term. When this acknowledgment comes, I believe questions of human improvement such as have been contemplated in prior chapters will be widely considered. I base these judgments on the general historical trend of humanity to discover and implement such changes in general circumstances as are considered desirable. The sort of equality, at the level of a well-off citizen of currently developed society, has widespread appeal. The urge to discover more about ourselves, so as to implement a hitherto undiscovered science of ethics will, I believe, have widespread appeal in a worldwide society of equals. If breakthroughs in life expectancy such as have been contemplated here are not yet accomplished, I believe there will be a push for such at that future time.

The correct ethical theory will explain how we should live, given prevailing circumstances. It will suggest the general manner in which those circumstances can be optimized for best living. It will likely indicate possible lines of evolution to better modes of living. These beliefs are speculative, and subject to falsification by various foreseeable and unforeseeable circumstances.

In this collection of writings I have previously cited ideas of Paul Churchland. I have been drawn to his writings through my interest in philosophical psychology. Of the various writers in this field, he is perhaps the most visionary. In his 1981 paper, "Eliminative Materialism and the Propositional Attitudes," Churchland speculates on "subsentential" modes of cognition which humans might adopt. His final speculation involves direct brain-to-brain communication of one person's "symphony of neural activity" to another.

> Once the channel is opened between two or more people, they can learn (*learn*) to exchange information and coordinate their behavior with the same intimacy and virtuosity displayed by your own cerebral hemispheres . . . If the entire population were thus fitted out, spoken language of any kind might well disappear completely, a victim of the "Why crawl when you can fly?" principle.[2]

I want to focus on this quotation both in itself and as marking a transition in the discussion. I want to lend the opposed view as much as possible, within the limitation of the overall space of considerations.

The opposed view, generally put, is that human beings are fine as they biologically are. Perhaps a few more years of average life expectancy would be a benefit. But parallel consciousness, sexlessness, indefinitely extended lifespan, and other such suggestions made in the preceding sections of this book, are to be avoided. The potential consequences are too extreme. Humanity, as currently biologically constituted, with little overall genetic variation for the past hundred thousand years or so, is an endpoint. Fine-tuning is in order in our sociopolitical relationships, greater psychological understanding of who we are would be welcome, but radical alteration is unnecessary and undesirable.

Whether Churchland thinks humans are in good standing as they are presently constituted, the quoted passage indicates a belief that improvement is both a fit subject of contemplation and, if possible, ought to be implemented. The immediate context of the quotation is a comparison of the immense data stream of communication (measured in bits per second) between the cerebral hemispheres as against that of ordinarily-spoken English. This comparison may be called into question, but accepting its validity for the while, the enormity of the difference signals a profound alteration of interpersonal communication.

It is expected that many people would demur on hearing Churchland's suggestion. I suspect among the typical responses to this suggested improvement would be the following: we are not in dire need of such a form of communication; we could not absorb this flood of information so as to continue our realtime activity; we enjoy life at our present rate of communication, and would enjoy life less if Churchland's contemplated increased rate were implemented.

To this list I add my criticism. Much as "a picture is worth a thousand words," so it might be that the "symphony of neural activity" is worth no more than a few words. That is, possibly in some useful but intuitive sense of "information" either no more, or insignificantly more, such information is transmitted at the neural level than at the simultaneously occurring spoken level. This observation bypasses discussion as to the desirability of the ostensible improvement were Churchland's suggestion implemented. It is more to Churchland's overall point of the dispensability of "the sentential level" of cognition.

I want to return to a previously mentioned criticism of Churchland's suggestion. Our verbal life is richly detailed. What would our life be like were we to replace words with direct, but nonverbal, neural communication? Suppose, for example, we could somehow speak intelligibly about the swarms of particles that, at one level, constitute our material-object surroundings? That is, suppose we could somehow in realtime be cognitively aware of these swarms, to the exclusion of our hitherto ordinary modes of awareness of medium-sized objects? Would our lives be better as appreciators of the various characteristics of these swarms, minus our former abilities? Previously, we noted the colors, textures, aromas, and so forth, of objects. Now, in our "swarm-awareness," we note average kinetic energy, density, distribution, and so forth.

One might object that these two modes of communication, ordinary and extraordinary, could coexist in the same individuals, much as was suggested for ordinary and extraordinary visual perception in "Mindful Seeing." Let us ignore this objection, for Churchland's eliminativism does not endorse it, and furthermore there are other suggestions in this book, such as sexlessness, that cannot coexist with our present ordinary condition.

I believe that many would rather live in the sensuous world of colors, textures, aromas, and so forth than what likely seems to them the sterile world of particle physics. For such people, knowledge of underlying causes, objects, and structures is not of sufficient worth to undertake large changes in human nature. This desire to live in the details of ordinary human life partly explains the disapproval of much twentieth-century painting by commonfolk. As Rembrandt and others have shown, there is much subtlety in the human face. To transform the human face into an African mask-like design, to fragment it, and ultimately to dispense with it except as a suggestion via title of the work-piece, is to lose such familiar subtlety. While inhabitants of the comparatively insular artworld have found gain in this loss, many others have not.

If all the suggestions in this book were implemented, human life would be sufficiently altered so as to merit the descriptive "post-human" occasionally appearing. Much of the detail of current life, much that occupies our minds, would be altered. One might argue that the reaction of the unschooled museum-goer to twentieth-century painting evinces the dislocation of the ordinary human faced with extraordinary possibilities for which they are unprepared. Those sensitized to the subtleties of the venue find much to appreciate and discuss.

Of course the connoisseur of modern art, the laboratory scientist, the theoretical physicist, and the microeconomist typically come home at the end of the day and live the ordinary life. Granted, some more and some less. For the most part, they do not lose touch with their common surroundings when disengaged from their specializations. This observation allows another manner of apprehending the overall suggestion of this book. What is being offered here is the transformation of our current ordinary lives (with the variability in ordinariness granted) to a radically different sort of ordinary life.

The contemplated sexlessness and solitariness in "Alone and Without Love" are key to what strikes the reader as a dismal vision. There are modern cities, such as Brasilia, in which the inhabitants are said to complain of isolation, boredom, and a general lack of the vitality of historically "organic" cities. In a word, cities such as Brasilia, and visions such as those presented here, seem *sterile*. Nations with planned economies, such as those of the former Soviet bloc, present a featurelessness and greyness to the outside observer. One trusts the human spirit to innovate based on awareness of immediate environment, an awareness that distant, long-range planning tends to dismiss.

It is noteworthy that Paul Churchland's vision of psychology as a neurally-based science, while leading him to suggest the color-drained subsentential scheme of communication cited above, provides some support for the ordinary human level of life as against my speculated post-human possibilities. The support to which I refer concerns the past two decades battle between the advocates of "classical" artificial intelligence (AI) programming and those of the connectionist persuasion. The former are represented in the philosophical community by Jerry Fodor and others, the latter by Churchland and others.

As is well known, classical AI has to this point failed to deliver on some of its famous promises, the most outstanding being Turing's prediction of what would be possible by the year 2000. While having success in areas such as chess-playing, medical diagnosis, and shortest-path finding, classical AI has not done well in areas such as pattern recognition and learning. Among the critical analyses of these shortcomings, the following are pertinent for this discussion: classical AI programming is overly rigid in its specification of alternatives; classical AI programming typically models a toy-world and then hopes to someday scale up; classical AI programming does not allow for a sufficiently exploratory and fault-tolerant learning that would allow for genuine being-in-the-world (rather, classical AI programming assumes full-blown knowledge with little room for learning). In short, classical AI idealizes the world in which it expects to operate, and in so doing it cannot cope with the richness and diversity of the actual world.

In contrast, the connectionist programming favored by Churchland allows for learning, flexibility, and is not in principle (at least in some of its manifestations) overwhelmed by vagueness, ambiguity, and diversity of possibilities. Viewed in this manner, such programming seems more true to life. While there are currently shortcomings to connectionist programming, its development over the last few decades has forced a re-evaluation of the promise of classical AI, particularly if the classical approach is unaided by connectionist "patches."

The comparison is hopefully not overly facile. Utopian schemes involving transformed human life are viewed here as similar to classical AI attempts to settle problems in a fixed, systematic manner. Humans have evolved from whatever earlier to the successful, adaptive, problem solvers that constitute homo sapiens sapiens. The life we humans lead is the life we have been suited by evolution to lead, and at an information-processing level our performance in leading such lives is best accounted for in terms of flexibility, adaptation, and learning, terms favored by proponents of connectionism.

I see in the preceding discussion the emergence of a dialectic of idealization opposed to detail. My utopian suggestions are of necessity idealized. Filling in the moment-by-moment lives of beings embodying these suggestions is impossible. Critics will claim that this impossibility is not simply due to the weight of such detailing, but also due to the lack of details to be supplied. How will a sexless, isolated, parallelly-conscious, etc., being occupy itself in an indefinitely extended lifespan? Any serious attempt to answer this question, the critic asserts, will force me back to detailing a life that makes sense for present-day humans.

This is a difficult criticism to deflect. Generally, the idealizations of physics are allowed, indeed respected, insofar as they apply to actual phenomena. In suggesting the supplanting of current humanity with significantly altered beings I cannot appeal to the attractiveness of these alterations to humans as presently constituted, most of whom would not favor various of my suggestions. These altered beings will likely be goal oriented in much the manner of present-day humans, which is to say that they will have long-term goals, short-term goals, and intermediate goals. These goals will generally differ somewhat from those of present humans. The post-humans I am considering will not have sexual goals, and will not seek the company of others in the manners in which we currently seek such company. Of course, goals are not necessary for being. Rocks, rivers, planets, and a host of other things do not have goals (purposes, desires). It is occasionally sensible to ascribe goals to machines, and not only machines which in some fashion are mimicking people. Whether such machine goals are intrinsic to the machine, rather than the human designer, is for the present questionable.

Having goals is not sufficient cause for present humans to endorse the suggestions offered in this work. Unfeeling machines can have goals. Alternatively, it might be argued that some variant of lotus-eating existence is goalless, except for the goal of lotus eating, yet is perhaps a desirable endpoint of human life. Hence, having goals might not be a necessary part of our contemplated future. Still, it would be difficult, if not impossible, to recommend a future in which humans or their descendants do not enjoy their lives, or are entirely bereft of what are commonly termed "feelings."

In recommending enhanced longevity, parallel consciousness, cognitively enhanced visual perception, sexlessness, and social isolation, I am not recommending life without enjoyment. Nor am I recommending lotus-eating purposelessness. I believe it possible to enjoy lives of increased cognitive powers, in which we appreciate the detail of the natural world as we learn more about it. Included in the natural world, as I intend the notion, is the biological world, the psychological world, and the sociohistorical world of humans and their descendants. As of now, much of our enjoyment in life is in living in the human world. Such living is not a theoretical-cognitive process, although it has cognitive aspects. It is my thought that as history progresses, we will de-emphasize many of the so-called "human" aspects of living, and take increasing enjoyment in the cognitive aspects. I do not mean that we will enrich our common, daily experience by becoming more cognizant of it. I do not mean that we will approach the aesthetics of our experience in a more critical mode. I do mean that, among many other subjects of thought, we will reflect on such experience as part of our past history, using concepts and theories that have yet to be discovered. We will become somewhat like what Plato had in mind for his Republic's philosophers, when left on their own on completion of their service.

I am aware how alien, how inhuman, these thoughts appear. I want to try once more to justify them through a consideration of the alternative.

The alternative is human life goes on as it has, either with gradual improvement or revolutionary change. That is, human life goes on for the better. There are other alternatives which are not in most people's judgment for the better (for example, the image Orwell's O'Brien offers Winston Smith regarding the society O'Brien's people are trying to create). I think this alternative will work, insofar as it does, for a limited future time. I invite the reader to consider two philosophical problems that indicate our limited future under this alternative.

There is the traditional philosophical problem of free will and determinism. Much has been written on this, and the discussion has become quite detailed and sophisticated. Yet there remains the nagging question, unproblematic only to the compatibilist: how are judgments of a person's moral responsibility justified if their behavior is metaphysically either unavoidable or not under their control? Suppose we are unsatisfied with compatibilist responses, unsatisfied with those responses that turn on the analysis of "can," and unsatisfied with the pragmatic/behaviorist response that holding people morally responsible conditions them to socially desired actions.

There is a tension here. Many do not want to think of themselves as determined beings. Yet libertarian positions may have no basis in fact. Some do not see the possibility of human society without holding its members genuinely morally responsible for their behavior. Yet human life goes on in the face of this philosophical difficulty, largely oblivious of the difficulty.

Assuming a general, worldwide uplifting of the educational level, there may come a time in which there is widespread recognition of this difficulty. For a long time humans generally acquiesced in the practice of slavery. This practice is no longer generally acceptable. Accepting the inconsistency of moral responsibility without genuine "contra-causal" freedom, we may eventually find such notions as personal autonomy and moral merit/blame inconsistent with what we know about people. If human life absent these notions becomes untenable in the manner in which human life with slavery became untenable, then humanity will be at a crisis. One way out of the crisis would be to change our pattern of living in a rather deep manner.

The preceding discussion will strike some as too "iffy." In any case, it is difficult for many to imagine the philosophical issue of freewill-determinism leading to a fundamental worldwide crisis. However, there are more pressing issues which are likely not avoidable. I choose the following such issue because of its close connection to the notion of worldwide equality discussed above. The issue is the conflict between humanity's hopes for equality and the widespread belief that such equality runs contrary to human nature.

The Marxist dream of equality has been waylaid by the collapse of various nominally Marxist regimes. Other forms of political economy pay occasional lip service to this ideal of equality, but often other aspects of their operating principles (such as free enterprise, marketplace value) work against it. There may come a time in the not-too-distant future, however, when people all over the world question economic disparities, inherited wealth, and private ownership of major means of production. Such questioning as exists now is for the most part kept under control by the propertied class.

No one knows whether the ideal of equality, where such equality is not a leveling down but rather an uplifting, will take hold among humanity so as to initiate worldwide change. It is not my intent here to preach the gospel of such change, nor do I wish to defend all that has transpired under the banner of Marxism, nor do I insist that it is only Marxist sociopolitical thought that allows for this ideal. Instead, I call attention to something many have noted. Many believe that genuine economic equality such as has been sketched above is either unrealizable, or if realized is inherently unstable. A primary reason supporting this belief is that such equality runs contrary to human nature. One often hears that people are not content to maintain equal status with their peers. Coupled with this belief is the assertion that the individualistic drive to better oneself, measured against one's peers, is often responsible for the innovations, intellectual and entrepreneurial, that have elevated living standards significantly beyond what they would have been were it not for such competitively-spurred innovation.

Some nominally Marxist regimes, such as Maoist China, have addressed the first part of this problem: attempting to create a society of equals. It is less clear that the goal was the "high-level" equality considered earlier in this chapter. Nor was it clear that Maoist China had solved the problem of fostering innovation while maintaining such equality. In any case, as with various social experiments in which generosity and comradeship are either presumed or thought capable of being instilled, Maoist China's genuinely revolutionary approach to daily living succumbed to the instability many of its critics predicted would occur upon the demise of its initial leadership. Perhaps had Mao sought a "higher level" equality, an educated and technologically fortified population might not have been so apparently receptive to the introduction of nonsocialistic "reforms." On the other hand, it might have been Mao's belief that such a population would instantiate the greed and selfishness he sought to expunge.

I do not think the problem of equality vs. competitiveness will long remain suppressed. Large segments of the world's population are currently underdeveloped, relative to Western standards of living. As more of these populations become literate, educated, better provided for, and more technologically endowed, those underdeveloped and impoverished people living in proximity to those more socially and materially fortunate can be expected to demand the sort of equality discussed here. If they do not receive this, one can expect all manner of discomposition. Yet fulfillment of such longing will not likely end the sort of competitiveness that exists among individuals and nations. Such competitiveness betrays the spirit of equality that often inspires many, and leads to systematic inequalities that exacerbate social tensions as they become entrenched.

The preceding consideration has been painted with a wide brush. Subtleties, qualifications, and rebuttals would require more discussion than is possible here, and the upshot would most probably remain inconclusive. As humanity becomes wiser and better adjusted to itself we may live as humans in peace and solidarity, if not friendship. I would not discount this possibility. Neither would I place full confidence in it. Which latter is to say that humanity may be doomed, although not necessarily to violent extinction.

One alternative to the violent extinction that we largely do not want is that we, in general agreement, undergo controlled evolution to something other than our present human form. I have given indication in earlier chapters as to what this might be. My speculations are intended to catalyze discussion. As such, I have advanced what I would term "bold" suggestions, which others might characterize less favorably. Perhaps a slight amount of "tweaking" is all that is necessary to overcome our inability to live peacefully with one another.

There is another reason, given in preceding chapters, for evolving from our present human state. This reason is largely inspired by Plato, who nevertheless does not typically speak of post-human evolution aside from various myths sprinkled through his dialogues. Our present psychological nature to some extent limits our cognitive potential. In simplified terms, assuming we could optimize our cognitive performance by filtering various distractions and distortions, we would approach the theoretical limit of such performance. One might consider such filtration an initial goal, if the aim is to do the best we can with what we have. If we desire to move beyond this "theoretical limit," however, then we must consider further alterations of our given capacities. Actually, to maximize filtration it might be necessary to alter our given nature in profound respects, even if our cognitive capacities remain much the same.

The tone of the above paragraph is admittedly dehumanizing. Partly, that is the point. Those who believe that ridding ourselves of various emotions and appetites leaves us with little or less, regardless of claimed cognitive benefits, will not be persuaded to follow this discussion further. They might be brought along with the observation that one permutation of the above suggestion would be to retain the noncognitive aspects of our present psychological makeup, yet increase our cognitive capacities over their present limitations. Those who object that our noncognitive psychological makeup is not fully separable from our cognitive abilities will favor this alternative only to the extent that they can be persuaded to pursue the sort of cognitive enhancement that is envisioned here.

Among other post-human futures, one might consider a Nietzschean vision of the overman as aesthetically creative. Assuming sufficient resources, such evolved subjects live as Michelangelo, Rembrandt, Beethoven, and similar humans. Perhaps they discover new forms of expression. They may appreciate each other's creations, or live more solitary lives. Their lives can be detailed within this wide indication in numerous manners. They may, for example, be sexual or nonsexual creatures.

One reason for preferring a post-human future of the cognitive sort is survival value. In the business of anticipating and responding to threats of extinction, science typically trumps art (with the proviso that science may create such threats). The Nietzschean-inspired future assumes sufficient resources, without an indication as to how such sufficiency is secured. One possible resolution of this opposition is to use science to secure a future in which the aesthetic life prevails. Given the uncertainties of any mode of being, however, it might prove disastrous to abandon science.

This brief consideration of Plato- and Nietzsche-inspired visions leads to another general possibility, favored in different specific manners by both these men. Suppose a distinction in types of evolved beings, so that some remain more or less human while others become beings with either the sort of capacities suggested in previous chapters, or with enhanced aesthetic sensibilities. As to the latter, even as we can attempt to imagine evolved beings as Einsteins and von Neumanns without excess human "baggage," so we try to imagine Mahlers and van Goghs without those emotional torments that detracted from their progress as creators (on the assumption that at least some of those torments hobbled, rather than enabled, their creative processes). All the while, we consider most of our descendants carrying on as now, somehow kept from their more destructive impulses (bearing in mind the likely increased means of destruction potentially, if not actually, available). A further possibility is a society of evolved beings some of whom are aesthetically hyper-endowed, others of whom are cognitively hyper-endowed.

Other possibilities will occur to those willing to make the effort. I have obviously sought to develop a vision of our descendants as cognizers. These hypothesized cognizers are not unfeeling beings, nor are they unaware of the prerequisites of their continued functioning as cognizers. The latter is necessary for their continuance, and the former for them to be more than merely sophisticated cognitive entities, without enjoyment of their accomplishments. Aside from their ability to continue the lives they want to lead, there is little I can argue in favor of their mode of being as against other modes that might be put forward. Moral arguments from the standpoint of our current humanity seem beside the point here, if such arguments are intended to legislate against our evolving into these beings. These speculated beings are not cruel or unconcerned with the welfare of any similar being that might need assistance, although it has been hypothesized in "Alone and Without Love" that such assistance will generally not be needed.

Moral arguments are applicable to the sorts of inequalities constituted by some of our descendants leading lives that others are incapable of leading, where the lives of the former are better than those of the latter, and where resources are available for all to lead the better lives. I realize the assumption of value objectivity made in the previous sentence is questionable. I believe some lives are objectively better than others, but for reasons given in "Facing Immortality" I cannot defend this belief in a manner ultimately satisfying to me. As has been noted, Plato's ideal society was unable to fulfill the condition of resource availability, and so he justified the diverse quality of lives by reason of the necessity of division of labor, the difference of intellectual capacity, and the understanding that all lead the best lives that social cohesion permits. Our speculated descendants will possess the means to allow all equal intellectual capacities. Withholding such will be morally unjustifiable. Further, even as we suspect that Plato's ideal society will fall victim to class warfare (as Plato himself imagines), so should we suspect that all developed societies with social inequalities and the material means (science, technology, and resources) of amelioration, have internal seeds of weakening, if not dissolution.

All this may be regarded as useless speculation. Humanity might relive an extended "dark ages." Barbarism has not vanished. More people are literate, but even the increasing number of those with so-called "higher education" does not translate into a well-informed, thoughtful populace. Political catchwords such as "democracy" and "free-market" are too often spoken in service of gain for the few. Or, if not a dismal future, humanity might discard the going assumption of progress through science and technology, choosing instead either stasis or regression (the latter of which I view as a dismal choice, but many do not). And so I plead guilty to the charge of extreme iffiness. My appeal is that what I have said concerning humanity's possibilities be considered in all seriousness. Regarding my more specific speculative predictions and proposals, I trust that proper skepticism will be exercised. I hope my optimism is recognized as such. I am saddened that I will not last to see if it is justified.

NOTES

1. Thomas Kuhn, *The Structure of Scientific Revolutions*. 2d ed. (Chicago: University of Chicago Press, 1970), 151.

2. Paul Churchland, *A Neurocomputational Perspective* (Cambridge, Mass.: MIT Press, 1989), 20–21.

Selected Bibliography

Aristotle. *Nichomachean Ethics*, trans. Martin Oswald. New York: Bobbs-Merrill, 1962.

Arnold, Mathew. *Poetry and Criticism of Mathew Arnold*, ed. A. Dwight Culler. Boston: Houghton Mifflin, 1961.

Bergson, Henri. *An Introduction to Metaphysics*, trans. T. E. Hulme. New York: Putnam, 1912.

Castaneda, Carlos. *A Separate Reality: Further Conversations with Don Juan*. New York: Simon and Schuster, 1972.

Churchland, Paul M. *A Neurocomputational Perspective*. Cambridge, Mass.: MIT Press, 1989.

———. *Scientific Realism and the Plasticity of Mind*. New York: Cambridge, 1979.

Cooney, Brian. *Posthumanity: Thinking Philosophically about the Future*. Lanham, Md.: Rowman & Littlefield, 2004.

Dennett, Daniel. "Where Am I?" in *Brainstorms: Philosophical Essays on Mind and Psychology*, by Daniel Dennett, 310–23. Cambridge, Mass.: MIT Press, 1985.

Descartes, Rene. *Discourse on Method and Meditations,* trans. Laurence Lafleur. Indianapolis: Bobbs-Merrill, 1960.

Dewey, John. *The Quest for Certainty*. Carbondale: Southern Illinois University Press, 1988.

Eddington, Arthur. *The Nature of the Physical World*. New York: Macmillan, 1928.

Fodor, Jerry, and Zenon Pylyshyn. "Connectionism and Cognitive Architecture: A Critical Analysis." *Cognition* 28, no.1–2 (March 1988): 3–71.

Freeman, Kathleen. *Ancilla to the Pre-Socratic Philosophers*. Cambridge, Mass.: Harvard University Press, 1956.

Fromherz, Peter, A. Offenhäusser, T. Vetter, and J. Weis. "A Neuron-Silicon Junction: A Retzius Cell of the Leech on An Insulated-Gate Field-Effect Transistor." *Science* 252 (1991): 1290–93.

Fukuyama, Francis. *Our Posthuman Future: Consequences of the Biotechnology Revolution*. New York: Farrar, Straus and Giroux, 2002.

Fuller, Steve. *Social Epistemology*. Bloomington: Indiana University Press, 1988.

Garreau, Joel. *Radical Evolution: The Promise and Peril of Enhancing Our Minds, Our Bodies – and What It Means to Be Human*. New York: Doubleday, 2005.

Gray, Chris. *Cyborg Citizen: Politics in the Posthuman Age*. New York: Routledge, 2001.

Halliday, David, and Robert Resnick. *Physics, Part I*. New York: John Wiley & Sons, 1966.

Haugeland, John. *Mind Design II*. Cambridge, Mass.: MIT Press, 1997.

Hume, David. *Dialogues Concerning Natural Religion*. Indianapolis, Ind.: Hackett, 1980.

Kant, Immanuel. *Critique of Practical Reason*, trans. Lewis White Beck. New York: Macmillan, 1993.

Kuhn, Thomas. *The Structure of Scientific Revolutions*. 2d ed. Chicago: University of Chicago Press, 1970.

Kurzweil, Ray. *The Age of Spiritual Machines: When Computers Exceed Human Intelligence*. New York: Viking Penguin, 1999.

Moravec, Hans. *Robot: Mere Machine to Transcendent Mind*. New York: Oxford University Press, 2000.

Mill, John Stuart. *Utilitarianism*. New York: Bobbs-Merrill, 1957.

Naam, Ramez. *More Than Human: Embracing the Promise of Biological Enhancement*. New York: Broadway Books, 2005.

Neisser, Ulric. *Cognitive Psychology*. New York: Appleton-Century-Crofts, 1967.

Nietzsche, Friedrich. *The Portable Nietzsche*, trans. and ed. Walter Kaufmann. New York: Viking, 1954.

Pashler, Harold. *The Psychology of Attention*. Cambridge, Mass.: MIT Press, 1998.

Plato, *Gorgias*, trans. Donald Zeyl. Indianapolis: Hackett, 1987.

——. *Phaedo*, trans. G.M.A. Grube. Indianapolis: Hackett, 1977.

——. *Protagoras*, trans. Stanley Lombardo and Karen Bell. Indianapolis: Hackett, 1992.

——. *Republic*, trans. G.M.A. Grube and C.D.C. Reeve. Indianapolis: Hackett, 1992.

——. *Sophist*, trans. Nicholas White. Indianapolis: Hackett, 1993.

——. *Symposium*, trans. Paul Woodruff and Alexander Nehamas. Indianapolis: Hackett, 1989.

Rorty, Richard. "The World Well Lost." *Journal of Philosophy* 69, no.19 (October 1972): 649–65.

Sartre, Jean-Paul. *Existentialism and Human Emotions*, trans. Bernard Frechtman and Hazel Barnes. New York: Citadel Press, 1957.

Schulz-Aellen, Marie-Françoise. *Aging and Human Longevity*. Boston: Birkhäuser, 1997.

Shostak, Stanley. *Becoming Immortal: Combining Cloning and Stem-Cell Therapy*. Albany: State University of New York Press, 2002.

Simon, Herbert. "Scientific Approaches to the Question of Consciousness." In *Scientific Approaches to Consciousness*, ed. Jonathan Cohen and Jonathan Schooler, 513–20. Mahwah, N.J.: Lawrence Erlbaum, 1997.

Snow, C. P. *The Two Cultures and the Scientific Revolution*. New York: Cambridge University Press, 1959.

Spinoza, Baruch. *The Ethics and Selected Letters*, trans. Samuel Shirley. Indianapolis: Hackett, 1982.

Stanovich, Keith. *The Robot's Rebellion*. Chicago: University of Chicago Press, 2004.

Trotsky, Leon. *Literature and Revolution*, trans. Rose Strunsky. New York: Russell and Russell, 1957.

Turing, Alan. "Computing Machinery and Intelligence." *Mind* 59, no. 236 (1950): 433–60.

Whorf, Benjamin Lee. *Language, Thought, and Reality: Selected Writings*. Cambridge, Mass.: MIT Press, 1956.

Young, Simon. *Designer Evolution: A Transhumanist Manifesto*. Amherst, N.Y.: Prometheus, 2006.

Index of Proper Names